River Morphodynamics and Stream Ecology of the Qinghai–Tibet Plateau

River Morphodynamics and Stream Ecology of the Qinghai–Tibet Plateau

Zhaoyin Wang

Department of Hydraulic Engineering, School of Civil Engineering, Tsinghua University, Beijing, China

Zhiwei Li

School of Hydraulic Engineering, Changsha University of Science & Technology, Changsha, Hunan, China

Mengzhen Xu

Department of Hydraulic Engineering, School of Civil Engineering, Tsinghua University, Beijing, China

Guoan Yu

Institute of Geographic Sciences and Natural Resources Research, Chinese Academy of Sciences, Beijing, China

CRC Press
Taylor & Francis Group
Boca Raton   London   New York   Leiden

CRC Press is an imprint of the
Taylor & Francis Group, an **informa** business

A BALKEMA BOOK

Cover photographs were taken by Zhiwei Li in May 2011 and September 2012, illustrating the Grand Canyon, incised bedrock channel, knickpoint reach, gravel braided channel of the Yarlung Tsangpo.

Published by: CRC Press/Balkema
P.O. Box 11320, 2301 EH Leiden, The Netherlands
e-mail: Pub.NL@taylorandfrancis.com
www.crcpress.com – www.taylorandfrancis.com

First issued in paperback 2020

CRC Press/Balkema is an imprint of the Taylor & Francis Group, an informa business
© 2016 Taylor & Francis Group, London, UK

Typeset by MPS Limited, Chennai, India

No claim to original U.S. Government works

ISBN 13: 978-0-367-57494-9 (pbk)
ISBN 13: 978-1-138-02771-8 (hbk)

Visit the Taylor & Francis Web site at
http://www.taylorandfrancis.com

and the CRC Press Web site at
http://www.crcpress.com

Library of Congress Cataloging-in-Publication Data

Table of contents

Preface

The Qinghai–Tibet Plateau, known as the roof of the world, is experiencing intensive tectonic motion and accelerating uplift because the Indian Plate moves northward and collides with the Eurasian Plate. The uplift of the plateau has been increasing the stream bed gradient resulting in substantial impacts on the evolution of the earth's surface, the fluvial processes, and the ecological environment. Evolution of rivers on the Plateau under this unique geological background has drawn significant attention from experts in different fields such as geology, river morphology, and ecology. The Yarlung Tsangpo (Brahmaputra River) and Sanjiangyuan (the sources of the Yellow, Yangtze (Jinsha) and Lancang (Mekong) rivers) are the most active and most important watersheds in terms of fluvial processes. In recent decades, climate change and human activities have changed the hydrology of most rivers on the plateau, reshaped the wetlands and lakes and affected the aquatic and terrestrial ecology. Studies on the plateau rivers are of fundamental importance for sustainable development and management of the rivers and protection of stream ecology.

Most areas on the plateau are very remote and difficult to access, which causes challenges for research on the streams. My research team performed field investigations 2–3 times every year since 2007. We have climbed over the Himalaya mountains and explored the Yarlung Tsangpo Grand Canyon. We took samples of water, sediment, and benthic invertebrates from the Yarlung Tsangpo and Palong Tsangpo rivers; measured the rate of erosion and sediment transportation by glaciers on Galongla Mountain; drilled sediment cores in the plateau deserts in Sanjiangyuan; and measured the stability of natural dams created by landslides, glaciers, and debris flows. All the data used in this book were collected from the plateau with elevations of 3,000–5,200 m after continuous painstaking efforts.

The main contents and authors are: Chapter 1 Geomorphology of the Qinghai-Tibet Plateau by Zhaoyin Wang, Le Liu, Guoan Yu, Xuzhao Wang, and Jun Du; Chapter 2 Fluvial Processes of Incised Rivers in Tibet by Guoan Yu, Lujie Han, Le Liu, and Zhaoyin Wang; Chapter 3 Meandering Rivers in Sanjiangyuan by Zhiwei Li; Chapter 4 Wetlands and Wetland Shrinkage by Zhiwei Li, Zhaoyin Wang, and Lujie Han; Chapter 5 Desertification and Restoration Strategies by Yanfu Li and Zhaoyin Wang; Chapter 6 Erosion and Vegetation by Zhaoyin Wang and Wenjing Shi; and Chapter 7 Aquatic Ecology by Mengzhen Xu and Baozhu Pan. Charles Steven Melching and/or Yiying Xiong polished the English of all chapters. Moreover, many other members of my research team have contributed in different ways to the creation of this book. They are: Xiaoping Xie, Lijian Qi, Kang Zhang, Chunzhen Wang, Haili Zhu,

Dandan Liu, Na Zhao, Wenzhe Li, Chendi Zhang, Xiaoli Fan, Xiongdong Zhou, and Liqun Lv.

This book is supported by the Ministry of Science and Technology of China (2011DFA20820, 2011DFG93160) and National Natural Science Foundation of China (51479091, 41571009, 91547112, 91547113, 51309154), Ministry of Water Resources of China (201501028), and Tsinghua Independent Research Program. Zhiwei Li and Guo-An Yu are supported by Changsha University of Science & Technology and Alexander von Humboldt Foundation, respectively. Although I have modified this book several times, there may be still some mistakes. I sincerely appreciate the readers' criticisms and suggestions for improvement.

Zhaoyin Wang
December 2015, Tsinghua University

Chapter 1

Geomorphology of the Qinghai–Tibet Plateau

1.1 UPLIFT OF THE QINGHAI–TIBET PLATEAU

According to the British Geological Survey, the Indian Plate moves northward at a rate of 50 mm/yr and collides with the Eurasian Plate, resulting in the uplift of the Himalaya Mountains and the Qinghai–Tibet Plateau (Zhang et al., 2004; Royden et al., 2008). The uplift of the plateau has been accompanied by the development of a series of large strike-slip faults and associated extensional normal faulting with average slip rates of 1–20 mm/yr (Tapponnier et al., 2001). Similar results have been obtained by other scientists (Yin and Harrison, 2000). Many researchers have reported that the tectonic motion and the uplift of the plateau are accelerating (Harvey and Wells, 1987; Coleman and Hodges, 1995; Chung et al., 1998). The Chinese Earthquake Bureau measured the current average rate of rise of the Himalaya Mountains at about 21 mm/yr.

The uplift of the plateau has been affecting the fluvial processes and reshaping the river morphology, particularly because the uplift rate of the plateau has been uneven along the Yarlung Tsangpo valley from west to east (TECAS, 1984; Hallet and Molnar, 2001; Li et al., 2006). The uplift of the plateau was the most important geological event during the Cenozoicera, forming a young geomorphological unit with the highest elevation and the largest mountainous area on earth. The uplift of the plateau has resulted in very important impacts on the evolution of the earth's surface, the fluvial processes, and the ecological environment. This uplift has drawn significant attention from experts in different fields such as geology, river morphology, ecology, and many others. The starting time of the uplift, its processes, rate, and mechanisms, and the impacts on the environment have been hot research topics on the plateau. Most areas on the plateau are very remote and difficult to access, which causes challenges for field investigations. Researchers have not reached consensus on the development processes of the geomorphology of the plateau. As research expands, the mysterious mask on the plateau has slowly been peeled off.

1.1.1 Geomorphological characteristics

The Qinghai–Tibet Plateau is located at the southwestern border of China, extending from the southern edge of the Himalaya Mountains to the northern edges of the Kunlun, Altun and Qilian mountains. The Pamir and Karakorum mountains are located on the west of the plateau. The Yulong SnowyMountain, Grand Snowy

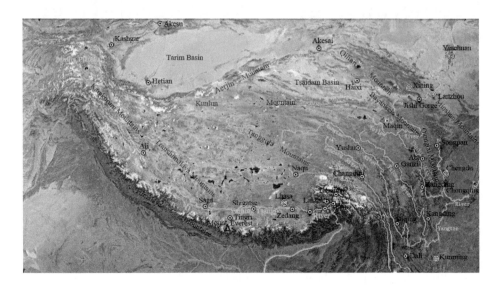

Figure 1.1 Map of the Qinghai–Tibet Plateau.

Mountain, Jiajin Mountain, Qionglai Mountain, and Min Mountain are located on the east. A map of the Plateau is shown in Figure 1.1.

The plateau covers an area of 3 million km², with 2.6 million km² within the territory of China, covering approximately 26.8% of the total land area in China. The area within China includes 11 cities and townships in 6 provinces and autonomous regions (i.e. Tibet and Xinjiang autonomous regions, and Qinghai, Sichuan, Gansu, and Yunnan provinces). Outside of China, the plateau covers areas in several countries including Bhutan, Nepal, India, Pakistan, Afghanistan, Tajikistan, and Kyrgyzstan. The name Qinghai–Tibet Plateau came from the fact that the main body of the plateau resides within the Tibet Autonomous Region and Qinghai Province.

The average elevation of the plateau is over 4500 m. It is also known as the roof of the world or the Third Pole. The topography on the plateau is rather flat with relatively continuous plains. The Qiangtang Prairie in the center of the plateau is classified as a high elevation hilly area with small relative elevation differences. The mountains surrounding the plateau have incision depths of 4000–5000 m (Li et al., 2006). Overall, the plateau is an intact plain surrounded by giant mountains and has a geographical characteristic of high on the west and low on the east. These features demonstrate that the whole plateau is an independent geomorphological unit.

The plateau has high elevation and low temperature. It is the largest glacial area in the low-latitude area of the earth. There are 36,793 glaciers on the plateau with a total area of 49,873 km² and an ice storage of 4,561 km³ (Liu et al., 2000). Using 0.86 kg/m³ as the density for ice, this storage of ice is equivalent to 3,923 billion m³ of fresh water, which is 10.8 times the total annual volume of runoff from the plateau. It contributes 50.4 billion m³ of glacial fresh water annually to rivers. The plateau is not only the water source for the Yangtze and Yellow rivers, but also the water source for the Yarlung Tsangpo (the upper reach of the Brahmaputra River), Lancang (the upper reach of the Mekong River), Nu (the upper reach of the Salween River), Shiquan (Indus), and Megawati Iloilo rivers. Therefore, it is called the Water Tower of Asia.

The plateau is still in its youth. So far, no ancient moraines older than 0.7 million years have been found on the plateau, which is another indication of its young age (Li, 2004). The main body of the plateau is the youngest part. At the end of the Tertiary period, strong tectonic lift and differential lifting movements caused the exterior drainage systems to disintegrate and then re-integrate around the subsidence basins, forming inland lakes and well-developed interior drainage systems.

The Hengduan Mountains with parallel hilly topography lie in the southeastern part of the plateau. Rivers in this area flow north to south through deep V-shaped valleys which lack broad floodplains and terraces. Broad valleys and large ancient plains exist in the areas above the highest knickpoints of the tributaries which have not been affected by headcut erosion. For instance, there are thousands of square kilometers of plains on the Haizi and Sulong mountains between the Jinsha (upper reach of the Yangtze River) and Yalong rivers.

There are many high mountains and deep valleys on the west of the plateau. The Pamir mountain range has the most rugged mountains and valleys and the deepest incisions on earth. For example, the Hongzha and Niubula rivers cut through not only the Himalaya Mountains but also the main ridge of the Karakorum Mountains due to headcut erosion. Through headcut erosion these rivers are currently extending towards the sources of many inland rivers in Asia. The elevation of the river mainstream in this area is usually below 2000 m, while the elevation of the main mountain ridge is as high as 7000 m, resulting in an elevation difference of 5000 m between the river and the mountain ridge. The bank slope is larger than 40° in many locations. These banks are extremely unstable. Large-scale landslides, avalanches, and debris flows often occur during rain storms. Many rivers in this region have U-shaped valleys with very large alluvial fans and alluvial plains, such as the Gilgit Valley near Gilgit and the Indian Valley near Djilas. The original plains which used to exist between the rivers have disappeared, indicating that the development of the geomorphology has reached the early stage of its prime life or late stage of its childhood.

1.1.2 Geological background

The Qinghai–Tibet Plateau is a relatively independent tectonic system in terms of the geophysical field and the lithospheric structure. The crust in the plateau area is twice as thick as that in other places around the earth, which is one of the most prominent characteristics of the plateau. According to the Bujia gravity anomaly map (Fig. 1.2) (Teng et al., 1980), a large area with negative gravity anomalies exists in the northern part of the Himalaya Mountains. The area with a high gradient of the gravity contours coincides with the edge of the plateau. The area with anomalies is almost identical to the horizontal location of the main plateau body where the crust is extremely thick. The thickness of the crust and the ground surface elevation are almost mirror symmetric. This indicates that the uplift of the plateau is closely related to the geological force that thickens the crust.

Gao et al. (2009) investigated the depth of the Moho surface on the plateau and its distribution characteristics based on results from three methods including deep

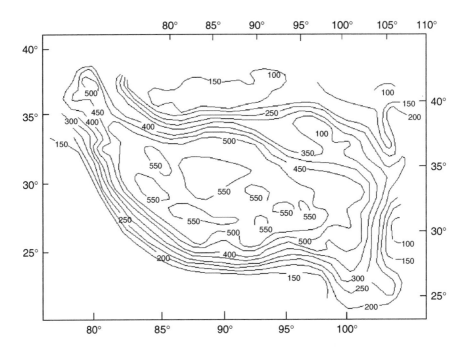

Figure 1.2 Bujia Gravity Anomaly (1°×1°) distribution of the Qinghai–Tibet Plateau (Teng et al., 1980).

seismic sounding, deep seismic reflection profile, and broadband digital seismograph. Gao's study found that the Moho surface of the plateau has a complex morphology and its depth varies drastically between different locations. The area with the largest depth was located at the west Kunlun tectonic knot and the depth was 90 ± 2 km. The shallowest area existed at the southern edge of the Ruoergai Basin with a depth of merely 49.5 km. The relative difference in the depth of the Moho surface was as large as 40 km. The depth also varied with location and tectonic unit, but, generally, it was deep in the west and south and shallow in the east and north.

The strata development of the plateau is rather complete. From old to new, the strata of the plateau are described as follows:

The oldest crystalline rock foundation was developed in the Archeozoic-Proterozoic eons. The stratum from the Archeozoic-Proterozoic eons is distributed in the northern part of the plateau, primarily consisting of moderately metamorphic rocks including gneiss, migmatite, and schist. The stratum also contains high-pressure and extra high-pressure metamorphic rocks including garnet lherzolite, eclogite, and granulite. The oldest stratum in the middle and southern parts of the plateau was developed in the Proterozoic eon, mainly consisting of moderately metamorphic rocks: gneiss, schist, leptynite, and marble. Some high-pressure metamorphic rocks, such as granulite, eclogite, and garnet amphibolite, exist in the Gandise-Himalaya area in the south.

The Paleozoic era stratum is widely distributed on the plateau with substantial diversity and a large amount of fossils. The Lower Paleozoic stratum is not fully developed, and mainly consists of mixtures of marine carbonates and clastic and phosphorus deposits from the Cambrian period. The Upper Paleozoic stratum consists of mixtures of marine-terrigenous rocks, marine carbonates rocks, and clastic rocks. In some areas, the stratum was already a terrestrial deposit during the Devonian period. Strata in the middle of the plateau mainly consist of mixtures of marine carbonate and clastic rocks. Strata in the Himalaya and Gandise-Tengchong areas in the south mainly consist of mixtures of marine clastic and carbonate rocks.

The Mesozoic era stratum is not fully developed in the northern part of the plateau, but is widely distributed in the middle and southern areas. Based on the characteristics of the sedimentary facies, it was determined that a marine regression process from north to south occurred during the Mesozoic era. By the end of the Late Cretaceous period, only some areas in the south still remained as oceans, while most areas in the middle and northern parts of the plateau had become dry land. The Mesozoic era stratum in the northern areas mainly consists of inland basin red clastic rocks. Only a few areas have Triassic period marine and marine-terrigenous clastic rocks with a small amount of carbonate rocks.

The Triassic period stratum in the middle of the plateau mainly consists of metastable to active marine clastic rocks with some carbonate rocks. The Late Triassic stratum mainly consists of mixtures of terrestrial and marine clastic and carbonate rocks.

The Jurassic period stratum is mainly distributed in the Qiangbei-Changdu area, Lanping, and the Qiangnan-Baoshan area and consists of marine and marine-terrigenous carbonate and clastic rocks.

The Cretaceous period stratum largely consists of terrestrial clastic deposits except for the west of the Qiangtang area, where the stratum consists of marine deposits. In the southern part of the plateau, the lower layer of the Mesozoic stratum is a mixture of marine clastic rocks and carbonates. The Late Triassic or Middle to Late Jurassic strata, especially the Late Cretaceous stratum unconformably contact the lower strata. The strata mainly consist of mixtures of marine and marine-terrigenous clastic and carbonate rocks.

The Cenozoic era stratum exists in the middle and northern parts of the plateau and mainly consists of inland basin deposits. The Cenozoic stratum does not exist widely on the plateau. The midwest area of the plateau still has some Eocenemarine-terrigenous sediment. The southern part of the plateau consists of inland basin terrestrial clastic rocks, except for the bottom layers of the Paleogene stratum in the Himalaya area where littoral neritic clastic and carbonate rocks still exist.

Magmatism on the plateau is very active and intense. The plateau is one of the regions with the most well-developed magmatic deposits in China. Many types of volcanic and intrusive rocks from the Proterozoic era to the Cenozoic era are exposed in this region. The exposed volcanic and intrusive rocks cover a total area of approximately $300,000\,km^2$, accounting for over 10% of the overall area of the region (Mo, 2011). The magmatic deposits from the Proterozoic era and the Early Paleozoic era are mainly distributed in the Qilian and Kunlun Mountains in the northern part of the plateau. The sizes of intrusive rocks are typically relatively small. They are typically in the forms of various sizes of granite, diorite stock, and basic to ultrabasic

Table 1.1 Formation periods of the mountains on the Plateau (Li, 1983).

Name of Mountain(s)	Highest marine stratum	Orogeny	Age of magmatic activity[1] (million years)
Altun Mountain	Devonian	Caledonian Orogeny	344–554
Kunlun Mountains	Lower Permian Lower Layer	Hercynian Orogeny	240–280
Tanggula Mountains	Middle Jurassic	Caledonian Orogeny	107–210
Mount Transhimalaya	Cretaceous	Yanshan Orogeny	30–79
Himalaya Mountains	Eocene Middle Layer	Himalaya Orogeny	10–20

[1]Determined using Krypton-Argon dating.

igneous rocks. Volcanic rocks are primarily andesite and basalt. Most magmatism began in the Late Paleozoic era. During this period, the coverage of the intrusive rocks expanded from the Qilian and Kunlun mountains in the northern part of the plateau to the balyanlkalla mountain-Songpan and Ganzi areas. Meso-acidic intrusive rocks and volcanic rocks are widely distributed throughout the entire region.

Magmatism during the Mesozoic and Cenozoic eras was very active, resulting in a continuous distribution of granite base rocks and large stocks. Large granite zones are concentrated in the Transhimalaya-Nyainqên Tanglha Mountain area. Meso-basic and meso-acidic volcanic rocks are also widely distributed in this areas. Magmatism occurred from north to south, which was consistent with the receding process of the ocean from north to south, both time- and space-wise.

Most areas on the Plateau used to be under the ocean between two paleo-continents. Due to the crustal movements, the ocean gradually receded from north to south. A mountain ridge was left behind after each recession period. The areas between the mountain ridges were relatively stable crust segments, featuring platform-type deposits and broad tectonics. Nowadays, these areas are basins between mountains and vast plateau plains.

Table 1.1 summarizes the highest marine stratum from north to south for each east-west mountain ridge on the Plateau, as well as the isotopic geological age of the major magmatism activities during the orogenic period. The information in Table 1.1 clearly demonstrates the process of orogeny and gradual recession of the ocean. The recession of the ocean laid the foundation for the uplift of the plateau.

1.1.3 Uplift process and mechanism

According to the evolution history of geology and morphology, the uplift of the Qinghai–Tibet Plateau is a complex process involving tectonic uplift, isostatic uplift, and the combined effects of weathering and erosion. Over the years, researchers have done many studies on the uplift of the plateau from perspectives of geomorphology, stratigraphy, meteorology, modern biology, paleontology, and petrology using methods such as paleo-magnetic measurements and isotopic dating. Many of these studies have shown that the uplift process is multi-staged, non-uniform, and non-constant. The rate of the uplift has changed from slow to rapid. The uplift has not yet stopped.

1.1.3.1   Initiation of the uplift

The uplift of the plateau is a result of the collision between the Indian Plate and the Eurasian Plate. This theory has been agreed upon by most researchers. Based on data obtained through paleo-magnetic measurements and isotopic dating, the Indian Plate had been drifting northward at a rate of 15–20 cm/yr since the end of the Late Cretaceous period (65 million years ago). In Middle Eocene epoch, the movement quickly slowed down to <10 cm/yr. The sudden decrease of the movement indicated the beginning of the collision of the two continental plates. Research by Li (2004) concluded that the initial collision of the Indian Plate and the Eurasian Plate followed the Yarlung Tsangpo River suture line.

The Late Eocene epoch sandstone above the Eocene nummulite limestone is still a marine stratum, which indicates that a portion of the ocean still existed in the area of the plateau during the Late Eocene epoch. The plateau entered into a slow uplift period after the ocean completely receded out of south Tibet. Mo (2011) studied the major collision zones that are located in south Tibet and over a distance of 1500 km, and concluded that the collision started 70/65 million years ago and completed approximately 45/40 million years ago.

All of the studies discussed in this subsection have proven that the collision occurred around 60 million years ago after the Indian Platehad drifted a long distance toward the Eurasian Plate, which started the uplift process of the plateau. At that time, the uplift was very weak and localized. Even after the collision of the two plates was completed in the late Eocene, the Indian Plate was still moving northward at a rate of 5 cm/yr (Li, 1995). Constant pushing and squeezing actions were resisted by the rigid land blocks on the north and east, forcing the crust in this area to shrink and thicken drastically. As a result, the plateau started to rise slowly.

1.1.3.2   Amplitude and stages of the uplift

Controversy still exists with regard to when the plateau reached its current elevation (Fig. 1.3). Popular theories by researchers outside of China are described in the following sections. Coleman and Hodges (1995), based on the fact that new-born minerals were found on the south-north normal fault of the Himalaya Mountains, concluded that the plateau had reached an elevation over 5000 m 14 million years ago and then subsided due to stretching in the east-west direction and its elevation dropped. Harrison et al. (1992) pointed out that the plateau started intense uplift in 20 million years ago and reached an elevation over 4000 m 14 million years ago. Rea (1992) stated that the plateau reached an elevation over 4000 m 17 million years ago and then subsided twice due to erosion, and then experienced drastic uplift again from a height of 1500 m 4 million years ago. Further, Tumer et al. (1993) concluded that the plateau had reached its current elevation 13 million years ago based on the age of the volcano in south Tibet. Spicer et al. (2003) concluded that the southern part of the plateau had reached its current height 15 million years ago based on the shape of fossils from plant leaves. Rowley and Currie (2006) measured the oxygen isotope ratio in the Lunbula basin on the plateau and concluded that the plateau had reached its current elevation 35 million years ago.

As a comprehensive scientific investigation of the plateau in the 1970s, Li et al. (1979) studied the stratified landforms on the plateau such as planation surfaces,

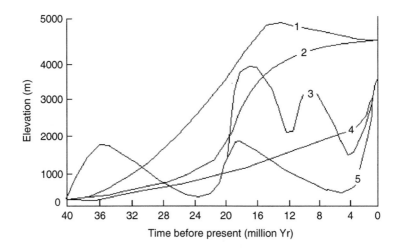

Figure 1.3 Various theories of the uplift process of the plateau (modified from Li and Fang, 1999) 1-Coleman and Hodges (1995); 2-Harrison et al. (1992); 3-Rea (1992) and Zhong and Lin (1995); 4-Xu et al. (1973); and 5-Li et al. (1996).

erosion surfaces, and river terraces. Based on the characteristics of these stratified landforms and measurements of their formation period, they brought up the theory of dividing the uplift process into "three uplift stages and two planation stages". During the time of the study, direct data for age measurement were not available, instead, the magnitude and staging of the uplift was determined based on indirect parameters such as palynomorphs, animal fossils, paleokarst, and ancient glaciers. However, these parameters can be significantly impacted by climate changes. Even though the results from these measurements were only relative, the general conclusions based on these measurements were still valid. At the end of the 20th century, geologists collected more reliable data and processed multiple indices and multidisciplinary verification methods. These studies improved the understanding of the "three uplift stages and two planation stages" theory as well as developed the theories of the Qinghai-Tibet Orogeny, Kunlun-Yellow Orogeny, and Gonghe Orogeny. These theories set up the framework of a complete theoretical system for the uplift of the plateau. The three uplift stages are described as follows.

The first stage of the uplift occurred in the mid to late Eocene epoch (approximately 40 million years ago) (Li, 1999). Mount Transhimalaya experienced the uplift first, however, the height of the uplift was no more than 2000 m, and the elevation of the mountainous areas in the northeast plateau was also only about 2000 m (Cui et al., 1996). After intense uplift movements, the plateau entered a relatively long and quiet period in terms of tectonic movements and started its first planation activity since the Cenozoic era. The planation surface formed during this period was the oldest planation surface on the plateau, formed during late Oligocene epoch to early Miocene epoch. It consisted of peneplains and piedmont denudation plains. The elevation at the time of formation was less than 500 m (Cui et al., 1996). The areas of the planation surfaces that still currently remain are relatively small, mostly distributed on top of large mountains. They are also called the Summit Planation Surfaces.

The second stage of the uplift occurred in the early Miocene epoch. Starting 25–23 million years ago (Pan, 2004), tectonic movements started expanding southward from Mount Transhimalaya. The majestic Himalaya Mountains rose up for the first time. The Summit Planation Surfaces started breaking up due to the tectonic uplift. When the uplift process paused, the plateau entered another long erosion planation period. It was the second planation period. The planation surface formed during this period is distributed widely on the plateau, and, therefore, is called the Main Planation Surface. The Main Planation Surface was formed 7.0–3.6 million years ago with a center elevation no more than 1000 m at the time of formation (Pan, 2004). Currently, the elevation of the Main Planation Surface is at about 4500 m (Cui et al., 1996), which and its surrounding mountains form the main body of the plateau. Based on the fact that most of the rivers on the plateau flow southward and eastward, it is obvious that the Main Planation Surface gently slopes towards the south and east. Before the formation of the Main Planation Surface the elevation was relatively low (<500 m) in the south and relatively high (1000–1500 m) in the north (between the Tanggula and Kunlun mountains). The middle part (between Mount Transhimalaya and the Tanggula Mountains) was a transitional zone (at an approximate elevation of 1000 m).

The third stage of the uplift began 3.6 million years ago (Li, 1999; Pan et al., 2004) and then the plateau entered into a new phase characterized by large-scale overall uplift. Li et al. (2004) named the three major tectonic movements during this period the Qinghai-Tibet Orogeny (divided into phases A, B, and C), Kunlun-Yellow Orogeny, and Gonghe Orogeny. Phase A of the Qinghai-Tibet Orogeny began 3.6 million years ago. The portion of the Main Planation Surface which had an average elevation of hundreds of meters (no more than 1000 m) started breaking up at a large scale due to the tectonic uplift. The thrust faults around the plateau experienced intensive activities, resulting in the accumulation of piedmont fan conglomerates During phase B of the Qinghai-Tibet Orogeny (2.6 million years ago), the plateau rose up to an elevation of 2000 m or so, and loess started accumulating. The East Asian winter monsoon occurred frequently during this period. During phase C of the Qinghai-Tibet Orogeny (1.7 million years ago), the ancient lakes on the Linxiadongshan lake disappeared, and the main trunk of the Yellow River was formed. The Qinghai-Tibet Orogeny formed the general shape of the plateau, however, the modern topography had not formed during this period. The Kunlun-Yellow Orogeny occurred between 1.2 and 0.6 million years ago, during which the Kunlun Mountains rose and the Yellow River cut through the Jishi Valley. Most parts of the plateau reached an elevation of 3000 m or higher, and a large number of the mountains reached the Cryosphere. The Gonghe Orogeny occurred during the early stage of the Late Pleistocene epoch. Since approximately 0.15 million years ago, the Yellow River incised through the Longyang Valley. The river incised 80–100 m in 100,000 years. The ancient lakes which existed in the basin since the Pleistocene epoch completely dried up and the plateau reached its current elevation.

Based on the foregoing descriptions, the uplift process of the plateau can be summarized as follows: The uplift began in the Eocene epoch, followed by planation in the Oligocene epoch. The second uplift occurred in the Miocene epoch, followed by another planation in the Pliocene epoch (Qin et al., 2013). Evidences for these two planation periods are the characteristics of the Summit Planation Surface and the Main Planation Surface. The third uplift occurred in the late Pliocene epoch, which went through phases A, B, and C of the Qinghai-Tibet Orogeny, the Kunlun-Yellow Orogeny, and the Gonghe Orogeny The plateau formed its current topography after the third stage of the uplift.

1.1.3.3 Rate of the uplift

During the long uplift process of the plateau, the rate of the uplift was constant. Xiao and Wang (1998) defined four stages of the rate change during the uplift process. They pointed out that the rate of the uplift has become higher and higher. Li et al. (2010) compiled a graph to describe the different phases of the uplift process of the plateau and its surrounding areas (Fig. 1.4), in which the viewpoints of Xiao and Wang was included.

I_1 Early Stage (60–50 million years ago) was a period with an extremely slow uplift rate of around 0.012–0.064 mm/yr;

I_2 Early Stage (50–25 million years ago) was a period with a slow uplift rate of around 0.07–0.31 mm/yr;

II_1 Mid Stage (25–11 million years ago) was a period with a medium uplift rate of around 0.13–0.62 mm/yr;

II_2 Mid Stage (10–3 million years ago) was a period with a medium uplift rate of around 0.30–2.05 mm/yr;

III Late Stage (2–0.5 million years ago) was a period with a rapid uplift rate of around 1.6–5.35 mm/yr;

IV Present Stage (since 0.5 million years ago) is a period with an extremely rapid uplift stage of 4.5 mm/yr (which can reach 10–14 mm/yr in the Himalaya Mountains).

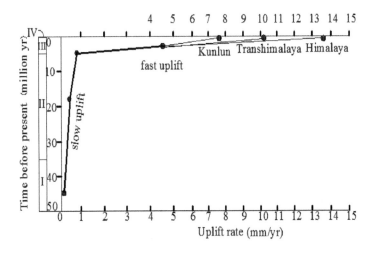

Figure 1.4 Rate of uplift since the Cenozoic era (modified from Li et al., 2010).

The uplift rate of the plateau is high in the south and low in the north. The squeezing effect from the Indian Plate on the Eurasian Plate is the main energy source for the uplift of the plateau. The extrusion energy at the beginning of the collision was primarily absorbed by the shrinkage of the crust. As a result, the uplift only occurred in localized areas. When the shrinkage of the crust reached its limit, the extrusion energy started getting released through thickening of the crust and uplift of the ground surface. During the uplift process, the shrinkage and thickening of the curst and uplift of the ground surface occurred in turns, and, at the same time, were affected by factors such as the gravity isostatic anomaly. As a result, the uplift of the plateau has been spatially unbalanced. The southern part of the plateau was closer to the collision zone and rose faster. The uplift rate got lower and lower for areas that are farther away from the collision zone towards the north. Current data show that the uplift rate in the Himalaya Mountains area varies from several mm/yr to teens of mm/yr since the Pliocene epoch, while it drops down to 0.42 mm/yr in the northern Altun Mountain area (SSBAF, 1992). This tendency is even more obvious in the directly measured data over the last several decades (Altun Fault Zone Research Group, China Earthquake Administration, 1992): the uplift rate is as high as 37 mm/yr at Mount Everest in the south; 10 mm/yr at Lhasa-Banda to the north; 8.9 mm/yr at Shiquanhe-Saga farther north; 6–9 mm/yr at Karakorum farther north; and finally drops down to no more than 5.2 mm/yr at Akesai at the northern edge of the Altun fault.

On the other hand, large differences exist in the uplift rate in even neighboring areas in the same region. Measured data collected by Global Positioning System (GPS) and precise leveling between 1966 and 1992 showed that the uplift of Mount Everest was extremely fast, reaching as high as 37 mm/yr, while the surrounding areas were only rising at 3.6–4.0 mm/yr (Xiao and Wang, 1998).

1.1.3.4   Mechanisms for uplift

Mechanisms for the uplift of the plateau have been a hot topic in the field of geology. Many theories have been developed by researchers from all over the world, such as the theory of mountain root floating, the subduction theory, the theory of even thickening of the lithosphere, the extrusion theory, the theory of multiple orogeny types and multiple uplift mechanisms, and so on. The understanding of the mechanisms for the uplift of the Plateau has improved as earth science advances and more modern tools become available.

Li (1999) pointed out that, based on the gravity anomaly (~500 milligal) and the 50~60 km-thick sial, it was easy to understand that the plateau reached its current elevation under isostatic compensation due to gravity. However, the real question was why there was such a thick sial on the plateau. Li concluded that the plateau rose because of the extremely thick sial, and the thickening of this layer was due to the intense squeezing action from both south and north. The horizontal compression came from changes in the autorotation velocity of the earth. Bending and folding of the crust due to horizontal compression was an important reason for the thickening of the curst, and the uplift of the plateau was a result of the isostatic effect from gravity.

Using the Plate Tectonic Theory to explain the mechanisms of the uplift of the plateau has been supported by evidence collected in geophysics, geochronology, and marine geology. Therefore, it is accepted by many geomorphologists. Results from pale-omagnetic measurements showed that the paleolatitude of the Inner Transhimalaya, Qiangtang, and Kunlun mountains on the plateau during the Permian period was 22.4°S, 16.1°S, and 11.9°S, respectively. Even though the current order of the loca-tions of the three terranes remain the same, their locations have moved northward by approximately 50°, and their relative locations have also adjusted (Dong et al., 1991; Li, 1995).

The period between the late Triassic and Jurassic periods was an important period for the north movement. During the Cretaceous period, the Transhimalaya and Qiangtang terranes started integrating with the stable Eurasian Plate on the north. In the Eocene epoch, the paleolatitude of Tingri in the Himalaya Mountains was 4.6°N, while its current latitude is 28.8°N, i.e. the Himalaya tectonic belt has moved north-ward by 24.2°, approximately 2,700 km, since the Eocene epoch. Another example is that the paleolatitude of Lhasa in the Eocene epoch was 13.8°N, which was 9.2° apart from Tingri. In other words the north-south distance from Lhasa to Tingri was 1,000 km. The current north south distance between the two cities is merely 120 km, meaning that the crust between the Himalaya tectonic belt and the Transhimalaya tectonic belt has shrunk by over 800 km.

Xiao and Wang (1998) studied materials in geology and geophysics, especially deep geophysics, and pointed out that the uplift of the Plateau is an uneven process affected by multiple factors, stages, and layers. The shrinkage, thickening, and uplift of the Plateau crust were controlled by three major energy sources. The first source was the squeezing action from the Indian Plate on the south and the resistance from the surrounding terranes such as the Tarim and Yangtze. The second source was the thermal effect inside the plateau, which not only increased the creep deformation and caused the crust to shrink and thicken, but also caused repeat thaw and thermal diffusion of the crust, creating a low-density space, and provided conditions for the floating and uplift of the crust. The third source was the controlling effect from the tectonic balance adjustment to the uplift. After the Pliocene epoch, except for the east and west corners where the squeezing effect was still relatively intense, the northward squeezing action from the Indian Plate weakened. It caused the 'subsidizing mountain root' to gradually rise up and result in the uplift.

Chen (1997) questioned the Plate Tectonics Theory. He conducted analyses on the crust structure, historical background, paleofloral zones, and the geothermal activi-ties on the plateau and concluded that the uplift was not directly associated with the continental collision. Chen believed that the uplift occurred as a result of a series of geological events: first the collision was caused by the mantle creep; then it experi-enced a 14 million year-long quiet period characterized by the formation of an intact Qinghai-Tibet paleo-platform; and then the uplift occurred in another active period. Mechanically, after the collision force and the follow up extrusion force subsidized; and after a stable period with vertical movements being the primary activity, then the uplift occurred as the crust experienced extrusion force from multiple directions again. The uplift was a product of another stress field under a separate crust evolution pro-cess. The uplift was caused by inland orogeny after the completion of the integration of the Indian Plate and the Eurasian Plate. The Indian Plate became a part of the Eurasian Plate, and the collision force and extrusion force was replaced by platform-type crustal movements.

1.2   IMPACTS OF THE UPLIFT OF THE PLATEAU ON STREAM NETWORKS

1.2.1   Stream ordering system and Horton's Laws

Dense river networks exist on the plateau. The Qinghai–Tibet plateau is the source of ten large rivers and is called the Water Tower of Asia. Since the uplift of the plateau occurs alongside the growth of the river networks, the uplift inevitably interferes with the growth of the river networks and affects their structures. River networks are water systems which have developed during the precipitation and erosion processes, and, therefore, follow certain topology laws. Horton (1945) developed a method of classifying and ordering the hierarchy of natural channels within a watershed. Strahler (1957) modified the method and the Horton-Strahler stream ordering system is probably the most popular system today.

The Horton-Strahler stream ordering system is illustrated in Figure 1.5. The uppermost channels in a drainage network (i.e., headwater channels with no upstream tributaries) are designated as first-order streams down to their first confluence. A second-order stream is formed below the confluence of two first-order channels. Third-order streams are created when two second-order channels join, and so on. In the stream ordering system, when two channels with different stream orders intersect, the stream below the intersection inherits the order of the stream with the higher order above the intersection (e.g., the stream below the intersection of a fourth-order stream and a second-order stream is still a fourth-order stream). Within a given drainage basin, the stream order correlates well with other basin parameters, such as the drainage area or channel length. Consequently, knowing the order of a stream can provide clues concerning other characteristics such as which longitudinal zone it resides in and the relative channel size and depth.

Horton's Laws: Horton (1945) pioneered quantitative studies on river morphology by introducing his laws of stream number, N_Ω, stream length, L_Ω, stream drainage area, A_Ω, and stream gradient, s_Ω; where Ω is the stream order, i.e.,

$$N_\Omega = Ae^{-B\Omega} \tag{1.1}$$

$$L_\Omega = Ce^{D\Omega} \tag{1.2}$$

$$s_\Omega = Ee^{-F\Omega} \tag{1.3}$$

$$A_\Omega = Ge^{H\Omega} \tag{1.4}$$

where N_Ω = number of Ω-th order streams; L_Ω = average length of the Ω-th order streams; s_Ω = average bed gradient of the Ω-th order streams; and A_Ω = average drainage area of the Ω-th order streams; and A, B, C, D, E, F, G, H are constants. The four equations are called the bifurcation law, length law, bed gradient law, and area law. These empirical laws were not derived from basic theories in physics or other fundamental theories. However, the validity of these equations has been independently verified and accepted as the basic laws in river morphology.

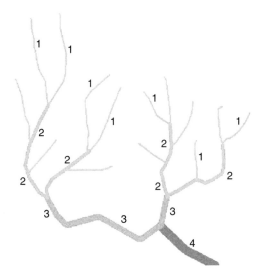

Figure 1.5 The Horton-Strahler stream ordering system.

The following equations were derived from Equations (1.1)–(1.4).

$$R_N = \frac{N_\Omega}{N_{\Omega+1}} = \frac{C_1 e^{-B_1\Omega}}{C_1 e^{-B_1(\Omega+1)}} = e^{B_1} \tag{1.5}$$

$$R_L = \frac{L_{\Omega+1}}{L_\Omega} = \frac{C_2 e^{B_2(\Omega+1)}}{C_2 e^{B_2\Omega}} = e^{B_2} \tag{1.6}$$

$$R_A = \frac{A_{\Omega+1}}{A_\Omega} = \frac{C_3 e^{B_3(\Omega+1)}}{C_3 e^{B_3\Omega}} = e^{B_3} \tag{1.7}$$

$$R_S = \frac{s_\Omega}{s_{\Omega+1}} = \frac{C_4 e^{-B_4\Omega}}{C_4 e^{-B_4(\Omega+1)}} = e^{B_4} \tag{1.8}$$

where R_B is called the Bifurcation Ratio which is defined as the ratio between the number of rivers in one order and the number of rivers in the next order; R_L is called the Length Ratio which is defined as the ratio of average river lengths between river orders; R_A is called the Area Ratio which is defined as the ratio of average drainage surface areas between river orders; and R_S is called the Gradient Ratio which is the ratio of average longitudinal gradients between river orders. The foregoing equations demonstrate that the values of R_B, R_L, R_A, and R_S are constants for all the rivers which follow Horton's Laws. These constants are called Horton's Ratios.

Many studies have shown that Horton's Ratios are almost the same for the natural rivers within the same river networks. Horton's Ratios are very close even for rivers in different river networks. For instance, the Bifurcation Ratio is usually around 4.5, and the Length Ratio and the Gradient Ratio are both around 2.0 (Smart, 1972; Liu and Wang, 2007). Therefore, some researchers believe that Horton's Laws reflect the "most likely condition" for any random river network. With the wide applications of

Fractal Science and Geographical Information Systems (GIS) in river network research, Horton's Laws have been used along with these two analysis tools to become a trend of river network research. Dodds and Rothman (2000) conducted studies on the sizes and self-similar relations of river networks, and explained many geomorphological phenomena based on their study results.

1.2.2 River networks on the Qinghai–Tibet Plateau

Geomorphology is the result of the long-term evolution of the earth surface under both internal and external forces. The uplift of the Qinghai–Tibet plateau impacts the formation of the geomorphology. For example, Gupta (1997) and Walcott and Michael (2009) conducted separate studies on the trellis stream network near the Himalaya Mountains and the conditions of the watershed outlets and concluded that these river systems were created by collisions of the plates.

The impacts of internal and external forces are particularly intense on the edge of the plateau. The uplift has resulted in the formation of many deeply incised valleys on the edge of the plateau. These deeply incised valleys are the most prominent geomorphological feature of the plateau edge. Due to the coupling relation between geomorphology and river network structures, the uplift of the plateau inevitably causes the river networks to deviate from the rules of normal river networks. During the long-term evolution process under the influences of both internal and external forces, six famous rivers have formed on the plateau. From north to south, they are the Yellow River, Yalong River, Jinsha River (i.e. the upper reach of the Yangtze River), Lancang River (the upper reach of the Mekong River), Nujiang River (the upper reach of the Salween River), and Yarlung Tsangpo River (the upper reach of the Brahmaputra River) (Fig. 1.6). All six rivers originate from mountains on the plateau at elevations above 4,500 m, flowing more than 1,000 km over the edge of the plateau, dropping over 3,000 m in elevation.

During the development of these rivers, the main body of the plateau was continuously rising. Under the strong influence of the geological uplift, the river networks on the plateau developed many unique characteristics compared to normal large river networks. Analyses were performed by the authors of this book to determine the Horton Ratios of the six stream networks. Based on data from the U.S. National Aeronautics and Space Agency Shuttle Radar Topography Mission (STRM) data and Digital Elevation Model (DEM) data from the China 1:250,000 topographic map database, modules were developed using the model-building function within ArcGIS. This tool automatically selected river networks and assigned orders to the six watersheds. In order to investigate the impacts of uplift of the plateau on the development of the river networks, the end points of the computations were all located in the river reach on the edge of the plateau. The number, length, area, gradient, and Horton Ratios were computed for different orders of rivers.

The hydrological analysis module within ArcGIS is also capable of building river networks, however, it involves many steps and does not have good solutions for handling false depressions. Since there are many deeply-incised valleys within the study area, blank values or errors often exist in the DEM data. Using the automatic filling function of ArcGIS would most likely generate false depressions and create parallel sub-order rivers, and, as a result, would affect the accuracy of the computations.

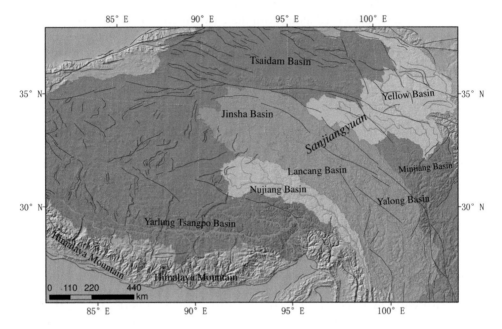

Figure 1.6 Six major river systems on the Qinghai–Tibet Plateau.

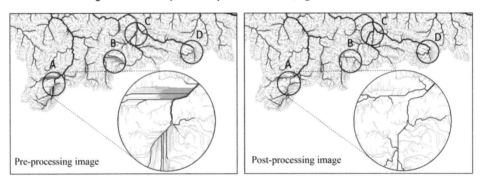

Figure 1.7 Process of depression handling and comparison of pre- and post-processing images.

The model-building function was utilized to create a module to avoid false depressions and build the river networks automatically. Figure 1.7 shows the general process for handling depressions and the final results. It can be seen that there are no parallel rivers in the processed DEM data after using the depression-handling module. After the river networks were established, the model-building function was used to create another module to assign orders and compute the river network characteristics and Horton Ratios.

Following processing of the DEM data, the small rivers within the six watersheds with drainage areas larger than 1 km² were identified using the modules previously described, stream orders were assigned, and the river networks characteristics and Horton Ratios were computed. The highest order the Yarlung Tsangpo watershed is 9

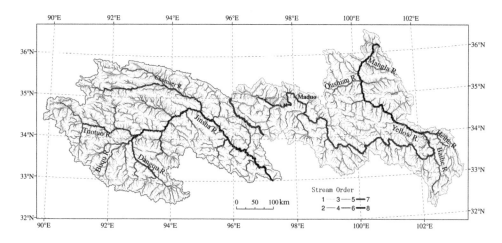

Figure 1.8 Stream networks of the sources of the Jinsha and Yellow rivers.

and the rest 5 river networks have the highest order of 8. The lower order rivers within the watersheds all mainly reside inside the plateau, while the highest order rivers all flow through the edge of the plateau. Fig. 1.8 shows the river networks at the sources of the Yangtze and Yellow rivers.

1.2.3 Influence of plateau uplift on topologic structure of the river networks

River networks on the plateau follow Horton's Laws, however, the topologic parameters on the edge of the plateau differ from typical river networks. Equations 1.5–1.8 were used to calculate the Horton Ratios. Figure 1.9 shows the relations between the stream order, Ω, and the average numbers of streams, N_Ω, the average length, L_Ω, and the average drainage area, A_Ω. It can be seen from the figure that these river networks generally follow Horton's Laws. Using logarithmic coordinate system, the relations between the river network parameters and the stream order are linear and can be depicted as straight lines. Most of these streams are located within the plateau. However, the highest order streams deviate from Horton's Laws, which implies that the uplift of the plateau affect the stream networks mainly at the edge of the plateau.

The Bifurcation Ratios were all found to be around 4.5, indicating that the river networks on the plateau all have topological features of typical dendritic stream networks (Liu and Wang, 2008). Similar results were also obtained for the Length Ratio, Area Ratio, and Gradient Ratio. Figure 1.10 shows that the Bifurcation Ratio, R_B, follows Horton's Law and remains a constant while the stream order Ω is lower than 7. However, the Bifurcation Ratio starts to deviate from the constant while Ω is equal to 7 or 8 and the value starts to fluctuate. The main cause for this condition is that the highest order rivers, while traveling through the edge of the plateau, have been influenced by geomorphological events such as faults and plate collisions. For instance, the highest order (i.e. 8) streams in the Lancang and Nujiang watersheds are very long, but the Bifurcation Ratios are relatively low, which indicates that the watersheds have

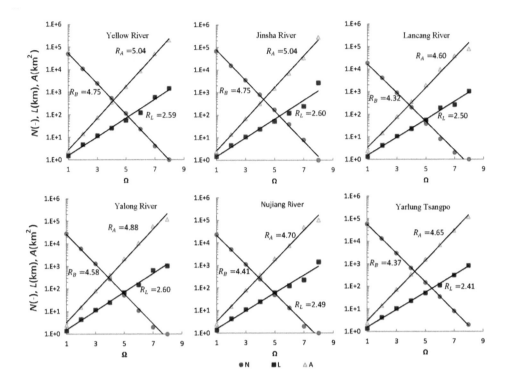

Figure 1.9 Relations between N_Ω, L_Ω, A_Ω, and Ω for the six major river networks.

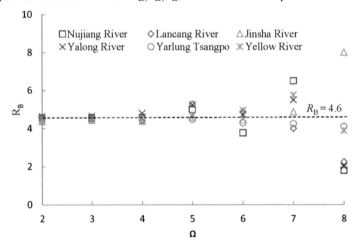

Figure 1.10 Relations between the bifurcation ratio R_B and stream order Ω.

become narrow due to the plate collision. Similar trends were also found for the Length Ratio, Area Ratio, and Gradient Ratio.

Figure 1.11 shows the relation between the stream order, Ω, and the average gradient, s_Ω for the six river networks. For lower stream orders the gradient decreases

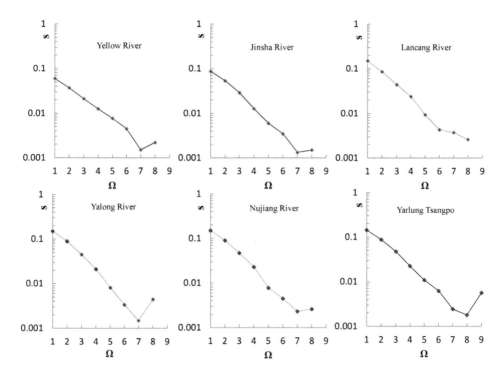

Figure 1.11   Relations between bed gradient s and stream order Ω.

with the order as indicated by Horton's Laws. However, for the highest stream order, the bed gradients increase as opposed to decrease. On the edge of the plateau, the uplift of the plateau has destroyed the typical river network patterns. The evolution of the geomorphology in these areas is progressing much slower than the uplift. Therefore, Horton's Laws are dramatically altered.

The Nujiang, Lancang, and Jinsha rivers flow in parallel on the plateau and form a geographical wonder. The narrowest stretch is less than 70 km wide, where the three rivers travel in parallel for 170 km without intercepting each other. The phenomenon is closely related to the uplift of the plateau. According to Ming (2006), all three rivers flowed southward prior to 3.4 million years ago. Since then, the transverse movement of the plateau has caused the development of the Hengduan Mountains and stopped the Jinsha River from flowing farther south. The Jinsha River then flowed eastward. Due to the lack of reliable deposition cross section and dating data, research on the evolution process of the three rivers has not been very successful. Only a few researchers have done studies on some of the rivers based on historical deposition materials (Ming, 2006; Zhang, 1998).

This book assumes that the three rivers used to be one large river. The Bifurcation Ratio of the combined river was calculated and compared to that of the three individual rivers in Table 1.2. According to Table 1.2, the Bifurcation Ratio of the combined river appeared to be more stable than the Bifurcation Ratios for the three individual rivers and generally followed Horton's Laws as for typical river networks. After the

Table 1.2 Stream number, N_Ω, and Bifurcation Ratio, R_B, for the top three orders of streams in the Jinsha, Lancang, and Nujiang river networks.

	Nujiang River		Lancang River		Jinsha River		Combined River	
Ω	N_Ω	$R_{B(\Omega-1)}$	N_Ω	$R_{B(\Omega-1)}$	N_Ω	$R_{B(\Omega-1)}$	N_Ω	$R_{B(\Omega-1)}$
6	13	3.8	8	4.9	39	4.8	60	3.9
7	2	6.5	2	4	8	4.9	12	5
8	1	2	1	2	1	8	3	4

combined river network was destroyed by the uplift of the plateau and divided into three separate river networks, the topological structure was also destroyed as a result and unique features started developing in the watersheds.

1.3   IMPACTS OF TECTONIC MOTION ON RIVER MORPHOLOGY

On the margin of the plateau, many rivers have experienced a long period of continuous bed incision because the uplift of the plateau has increased the stream bed gradient (Lavé and Avouac, 2001). This is in contrast with the long-term sediment deposition in the wide river valley along the Yarlung Tsangpo River where an extremely thick layer of sediment has accumulated. This fact demonstrates that the uplift in the Himalaya Mountains and the plateau has been uneven and the river reaches with a lower uplift rate have been storing sediment. Figure 1.12 shows the distribution of the fault zones on the plateau and the northward movement and their uplift rates. The Himalaya Mountains are moving northward at approximately 50 mm/yr and the uplift rate is approximately 21 mm/yr. At the Tanggula Mountains, the northward movement rate is about 40 mm/yr and the uplift rate is about 15 mm/yr. Other than the large east-west fault lines, there are six longitudinal fault lines across the Yarlung Tsangpo and the Y-shaped fault lines of Xianshuihe-Lungmenshan and the north-west Karakorum fault line.

Due to the collision between the Indian Plate and Eurasian Plate and the uneven uplift, the six fault lines were stretched and fractured and cut through the Himalaya Valley. For example, a fault line developed from the upstream end of the Tuoxia-Yongda valley and cut through the Himalaya Valley (between CS6 and CS7 in Figure 1.14). The east part of the fault line moved higher than the west part, which was caused by uneven uplift rates. Uneven uplift has caused several grabens (a depressed block of land bordered by parallel faults). Sediment has slowly accumulated in these grabens. This is a very slow and long process, hence, the stored sediment has kept the composition features of mountainous river deposits. Figure 1.13 shows the impacts of uneven uplift of the Himalaya Mountains on the development of wide valleys and gorges. A large amount of sand and gravel has deposited along the graben reaches, resulting in wide valleys with braided and winding river channels (Wang et al., 2015).

1.3.1   Development of the Yarlung Tsangpo River

The uplift of the plateau essentially created the macro-relief of the Yarlung Tsangpo basin. After the formation of the stream network, the continuous tectonic motion and

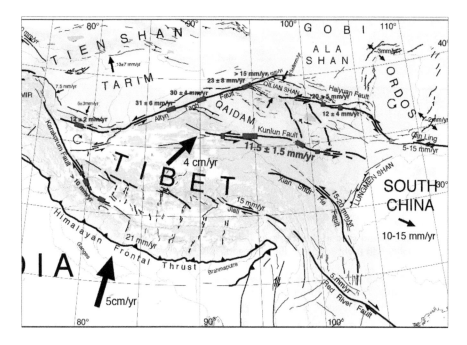

Figure 1.12 Distribution of fault lines on the Qinghai–Tibet Plateau and uplift rates of different sections.

Figure 1.13 Gravel and sand were continuously stored in a graben, which resulted from the uneven uplift, and the river valley becomes very wide (left). Incision occurred in the reaches with high uplift rates, which resulted in a narrow and V-shaped valley covered with a thin layer of boulders and cobbles (right).

uplift of the plateau have been affecting the fluvial process and reshaping the river morphology, particularly because the uplift rate of the plateau has been uneven along the Yarlung Tsangpo valley from west to east (Hallet and Molnar, 2001; Li et al., 2006).

Figure 1.14 Gravel and sand in the top layer (left) and sediment core about 300 m underneath the valley bed (right) of the Yarlung Tsangpo River.

The Yarlung Tsangpo River (Tsangpo means river in Tibetan) is the upper reach of the Brahmaputra River. The river originates from the Jimayangzong Glacier on the northern slope of the Himalaya Mountains and flows eastward for about 1,600 km before pouring into the Grand Canyon of the Yarlung Tsangpo, which is located from Pai to Pasighat (Fig. 1.15). Different from any other river, the fluvial process of the Yarlung Tsangpo River occurs concurrently with the uplift of the Himalaya Mountains. Therefore, the river exhibits unique features in morphology and sediment movement (Wang and Zhang, 2012). Although a large amount of suspended sediment such as fine sand and silt travel downstream into the Brahmaputra River during floods, most pebbles and cobbles have settled along the low gradient river reaches and formed a thick deposition layer. The river consists of alternating sections of wide valleys (i.e. low gradient river reaches) and gorges (i.e. high gradient river reaches). The wide valley sections have braided and anastomosing channels, gentle bed gradients, and thick alluvial deposits. In contrast, channel bed incision has occurred in the high gradient reaches (gorge sections), where a high rate of uplift has occurred. Thus, the gorge sections exhibit single, straight and deeply incised channels with steep gradients, rock channel beds, and several terraces (TECAS, 1983, 1984; Hallet and Molnar, 2001; Zhang, 1998).

A huge amount of sediment is deposited in the wide valley sections, which consists of gravel and sand. Core samples were taken in recent geological investigations from 10 wells with depths varying from 140–600 m. In three wide valley sections with a width of about 1,500 m, the samples from the bed surface and depths of 120, 140, 105, and 685 m consist mainly of gravel mixed with sand in the interstices of the gravel as shown in Figure 1.14 (He et al., 2008, Wang and his team, 2015). In the gorge sections, however, the channel bed consists of bed rock. Only a shallow layer of boulders covers the rocky bed.

Wang et al. (2015) selected twenty-nine cross sections and detected the bed rock depth and sediment deposition along a 1,000-km-long reach of the Yarlung Tsangpo valley from Xietongmen to the Yarlung Tsangpo Canyon from 2009 to 2012. The

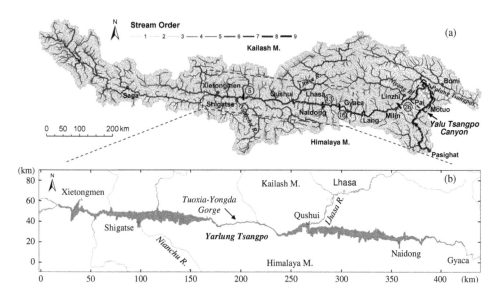

Figure 1.15 (a) Stream network in the Yarlung Tsangpo basin; (b) Plan view of the Yarlung Tsangpo River with gorge sections (200 m wide) between wide valleys (10 km wide).

STRATAGEM EH4 electromagnetic imaging system jointly produced by Geometrics, Inc. and Electromagnetic Instruments, Inc. (EMI), was used to measure the depth of the interface of bed rock and the sediment deposits. The depth of the sediment deposits can be detected by this instrument with a sharp change of the electromagnetic signal at the interface. The detected depth of sediment deposits measured with the STRATAGEM EH4 was compared with the data from core borings. The instrument was found to have a relative error of less than 10%. In the field investigations, several shallow deposit profiles were found at gravel mining sites and highway and bridge construction sites, where deposit samples were taken, or the depth of the bed rock surface was estimated.

A contour laser range meter made by Kustom Signals Inc., with an accuracy of 0.1 m and a measurement range of 2,000 m was used for distance measurement from the banks along the cross sections. A meridian color GPS made by Magellan was used to measure the elevations and positions. This measurement system has a maximum error of 1 m. SRTM3 (resolution 90 m) from the U.S. NASA provided the basic DEM terrain data. ArcGIS was used to extract longitudinal profiles and river channel gradients. Google Earth satellite images were used to examine planform features of fluvial morphology and river patterns. The flow discharge, velocity, sediment concentration, and suspended sediment load data at Daqiao (Lhasa), Nugsha (88 km downstream of Shigatse), and Nuxia (Linzhi) stations were collected in the 1960s, 1980s, and 1990s. The sediment deposits were sampled from depths of 0.5–4 m in the wide valley sections. Size distributions of sediment were obtained by sieving the samples.

Figure 1.15a shows the stream network of the Yarlung Tsangpo River. The drainage area is elongated in the east-west direction and confined in the north-south direction because the northward movement of the Indian Plate squeezed the basin and confined

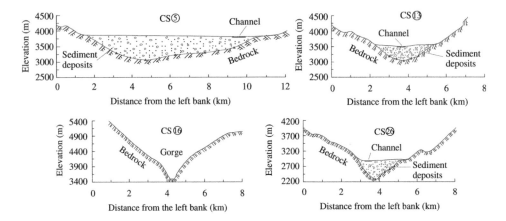

Figure 1.16   Four cross sections showing the present valley bed and the interface between the sediment deposits and bed rock surface: CS 5 – downstream of Shigtase; CS 13 – Naidong; CS 16 – Zengga Gorge; and CS 26 – Milin.

its development in the north and south directions. The stream network consists of 57,100 first-order streams and the river becomes a 9th-order stream at the Grand Canyon of the Yarlung Tsangpo below the confluence with the Palung Tsangpo River. The studied reach of the river is from Xietongmen to the Palung Tsangpo confluence with the Yarlung Tsangpo, basically the upper one fourth part of the Yarlung Tsangpo Canyon.

In the plan view, the Yarlung Tsangpo valley looks like lotus roots with gorge sections with widths of only 0.1–0.2 km between wide valleys with widths of 3–10 km, as shown in Figure 1.15b. The brown area in Figure 1.15b is the valley defined by the boundary of the sediment deposits and the mountains, while the blue curves are the river channels, which are a single thread in the gorge sections but multiple threads in the wide valleys. Numerous bars and islands are formed in the wide valley sections. In the gorge sections, bed rock is exposed to the water flow at many places and big boulders, 1 m in size or larger, overlap and form step-pools in the channel. Several terraces of 10–30 m high are on one side of the two banks, which is an indication of a long period of river bed incision (Wang et al., 2014). In the wide valleys, however, there is a very thick layer of sediment deposits consisting of gravel and sand, which indicates long-term aggradation.

Figure 1.16 shows three cross sections in the wide valleys and one cross section in a gorge section. The depth of the bed rock surface or the interface between the bed rock and the sediment deposit, and the thickness of the sediment deposit layer are also shown in the figure. The locations of the four cross sections in the plan view and the longitudinal profile are shown in Figure 1.15a and 1.17.

The original river valley at the cross sections was V-shaped. Long term sediment deposition caused aggradation and resulted in flat and wide valleys, as shown for Cross Sections (CS) 5, 13, and 26 in Figure 1.16. The present channel is developing on the sediment deposits, and is not stable, thus, migrating within the wide valleys. The river pattern in the wide valleys is braided. The floodplain is large and flat with very poor

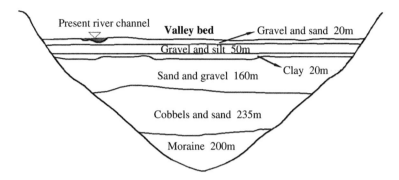

Figure 1.17 Thickness of sediment deposit layers and present valley bed surface of the Yarlung Tsangpo at Milin.

vegetation. The valleys are as wide as 10 km. At the gorge sections such as CS 16, the valley remains V-shaped and the river pattern is straight. Because the bed rock channel has experienced rapid incision, the lower part of the bank slope is steeper than the upper part, which forms an super V-shaped valley (Wang et al., 2014). There is almost no sediment deposition in the gorge section, where the bed gradient and flow velocity are high. The bed rock surface is close to the channel bed and the bed rock in several places is exposed to the water flow.

Figure 1.17 shows the cross section of the sediment deposit at Milin, which was obtained from the cores of drilling wells. The depth of the deposit is 685 m, which is exactly the same as the measured value with the instrument EH4. The original cross section of the valley was V-shaped. Sedimentation changed the valley into U-shape. The present valley bed is quite flat and the present river channel occupies a small portion of the valley at the left side of the valley (Wang and his team, 2015).

1.3.2   Massive sediment storage in the Yarlung Tsangpo

Figure 1.18 shows the longitudinal profile of the current channel thalweg, the profile of the interface between the sediment deposit and bed rock, and the mean mountain elevation within 10 km of the river valley in the lateral direction from the river channel. The profiles of the thalweg and interface were obtained by connecting the lowest points of the current channel bed and the sediment-rock interface for the 29 measured cross sections, respectively. The points A, B, and C and Milin in the figure are the locations of the core borings. The solid triangles are the depth of the sediment deposit measured with the Stratagem EH4. The hollow triangles are the depth of the sediment deposit calculated based on the interface elevation of the exposed rock on the channel bed and the core boring data. The bed rock surface in the gorge sections is several hundred meters higher than the interface in the upstream wide valleys, which forms several massive "sediment reservoirs". Gravel and sand have been filling these "reservoirs" over the past millions of years, over which a braided fluvial river has developed. In the gorge sections, the bed gradient is high and the water flow has scoured the rocky bed, which has resulted in long term stream bed incision. These gorge sections have become

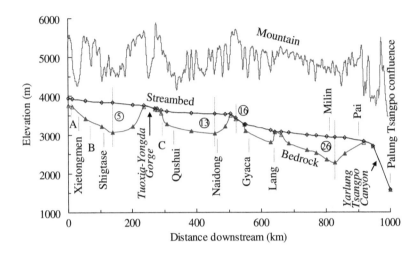

Figure 1.18 Longitudinal profiles of the thalweg of the Yarlung Tsangpo, the interface between the bed
rock and sediment deposits, and the bank mountain elevation.

knickpoints. The three knickpoints near Saga (located upstream of Xietongmen and
out of the range of Fig. 1.15b), Gyaca, and Pai were believed to indicate three stages
of the uplift in the Himalayan Movements in the Quaternary period (TECAS, 1983).

On the slope of the Himalaya Mountains, fluvial incision occurred in the rivers
following the uplift of the mountains (Burbank et al., 1996; Lavé and Avouac, 2001).
Nevertheless, a long period of sediment deposition occurred in the Yarlung Tsangpo
valley and a massive amount of sediment has been stored, which is a very unique geomorphological condition. The river is on the plateau, and the uplift of the plateau does
not directly increase the river bed gradient except for the Yarlung Tsangpo Canyon,
which is on the south slope of the plateau. Many tension faults are present across the
river valley. For instance, a fault across the river valley occurred at the upper end of
the Tuoxia-Yongda Gorge between CS 7 and CS 8. The east side of the fault moved
upwards relative to the west side. The relative movement is due to the differing uplift
rates. A high rate of uplift occurred in some sections and a lower rate of uplift occurred
in other sections along the river valley. The uneven uplift resulted in grabens and sediment deposition in these grabens. It has been proved that the three knickpoints at
Saga, Gyaca and Pai represent three uplift stages of the Himalaya (TSIT-CAS, 1983).

Figure 1.19 shows the maximum depth of the sediment deposit and the width of
the valley along the Yarlung Tsangpo River from Xietongmen to the Yarlung Tsangpo
Canyon. The large width of the valley corresponds to the large depth of the sediment
deposits. This is because the sediment buried the valley and the narrow V-shaped
channel has been changed into a wide U-shaped valley. Table 1.3 lists the maximum
depth of the sediment deposits and sedimentation volume (V) between the neighbor
cross sections. The thickness of the sediment deposits (H_m) is 400–800 m. For each
cross section the area of the sediment deposits between the valley bed surface and the
sediment-rock interface was calculated. The total volume of the sediment deposit in the
1,000 km long river valley was obtained by summing the products of the cross sectional

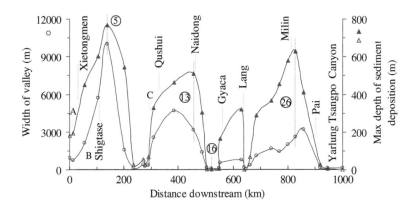

Figure 1.19 Depth of sediment deposits and valley width along the Yarlung Tsangpo River.

areas of sediment deposit and the distances between neighboring cross sections. The total volume of sediment storage in the river valley was calculated to be about 516 billion m^3.

The average annual water runoff in the Yarlung Tsangpo River from 1956–2000 at Nuxia (Linzhi) was 60.6 billion m^3 and the annual suspended sediment load (mainly fine sand) was about 30 million tons. The bed load has not been measured in the Yarlung Tsangpo River, however, it was estimated based on bed load data in the Yangtze River. The bed load transportation in the Yangtze River was measured for the Three Gorges Project (Wang et al., 2007). According to these measurements, the average annual bed load at the upstream end of the Three Gorges Reservoir was about 0.4 million m^3, which consists mainly of gravel and coarse sand (Han, 2009). The bed load transported from upstream of the Jinsha River has been trapped by the massive Hutiaoxia Gorge landslide dam for several centuries. Therefore, the measured bed load on the Yangtze River was produced in the reach from the Hutiaoxia Gorge to the upper end of the Three Gorges Reservoir, which has similar reach length and effective erosion area as the Yarlung Tsangpo River. Assuming the same annual sediment storage rate, the total volume of sediment stored in the Yarlung Tsangpo River was estimated to be 516 billion m^3 during a period of 1.2 million years (Wang et al., 2015). In other words, the river continuously stored sediment during the uneven uplift of the plateau over the last millions of years. Up to now, a massive amount of sediment has been stored in the wide river valley sections, which created 5,000 km^2 of plains with maximum depths of 350–800 m of sediment.

The Tuoxia-Yongda, upper Gyaca, Lang, and Yarlung Tsangpo Canyon sections have been rising faster than the Shigatse, Qushui-Naidong, lower Gyaca, and Milin sections of the Yarlung Tsangpo River, as shown in Figure 1.18. Fluvial processes have concurrently occurred with the geological processes. Sediment has continuously deposited in the slowly-rising sections. In the meantime, channel bed incision has occurred in the fast-rising sections. During a long evolution process, the rock bed in the river reach in the Shigatse, Qushui-Naidong, Gyaca, and Milin sections has been covered by deep sediment deposits and the river valley became very wide. On the other

Table 1.3 Volume of sediment storage between the cross sections of the Yarlung Tsangpo valley.

CS	L (km)[a]	S (km)[b]	W (m)[c]	H_m (m)[d]	V (10^9 m³)[e]	Note
1	0.00	0.00	730	193	9.497	
2	40.76	40.76	2170	450	53.043	
3	50.77	91.53	5760	604	90.043	
4	33.23	124.76	10040	768	120.695	
5	63.28	188.04	1590	544	4.198	
6	33.91	221.95	140	28	0.202	Tuoxia-Yongda Gorge
7	35.70	257.65	400	55	0.061	
8	3.10	260.75	1520	27	0.071	
9	3.29	264.04	260	70	0.111	
10	13.10	277.14	390	35	2.334	
11	17.21	294.35	2570	332	54.932	
12	75.58	369.93	4730	465	70.333	
13	73.05	442.98	3170	510	14.678	
14	31.47	474.45	1400	307	1.081	
15	16.31	490.76	270	11	0.049	Upper Gyaca Gorge
16	30.54	521.30	106	3	0.243	
17	12.53	533.83	240	7	0.082	
18	4.74	538.57	560	166	6.419	
19	77.31	615.88	804	321	0.488	
20	11.83	627.71	208	5	0.062	Lang Gorge
21	20.62	648.33	394	35	1.433	
22	23.06	671.39	1140	289	12.650	
23	54.86	726.25	1670	368	9.312	
24	28.92	755.17	1460	456	14.512	
25	32.56	787.73	1990	578	18.442	
26	26.55	814.28	2610	630	22.117	
27	29.24	843.52	3200	412	11.181	
28	60.28	903.80	167	29	0.038	Yarlung Tsangpo Canyon
29	29.03	932.83	139	5	0.011	
30	53.57	986.40	130	1		
Sum					**518.3**	

a) Distance between cross sections; b) Accumulated distance from cross section 1 (CS 1); c) Width of valley bottom;
d) Maxium depth of sediment deposition of the cross section; e) Volume of sediment storage between cross sections.

hand, the river reach in the Tuoxia-Yongda, upper Gyaca, and Lang sections turned into a deeply incised river valley and the reach below Pai became a several kilometers deep grand canyon.

Figure 1.20 shows the size distributions (b) and pictures of sediment deposits at the wide valley section, CS5 (a), and Gyaca gorge section, CS16 (c). The size distributions of the sediment deposits at the two wide valley sections (CS5 and CS13) are almost the same. The sediment consists of gravel (about 75%) and sand (about 25%). Samples taken from core borings and shallow sediment profiles at sediment mining sites showed that the sediment composition differs little at different depths and locations in the wide valley sections. This fact implies that the difference in the uplift rate occurred over a very long period of time, which formed high gradient river reaches and low gradient reaches rather than "dams" and "lakes". A small portion of the gravel might have transported downstream in the river channel, but most of the gravel has deposited in

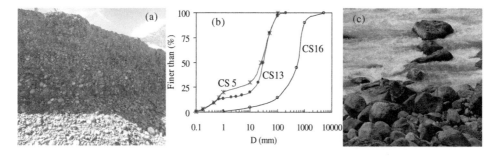

Figure 1.20 (a) A portion of the 700 m thick layer of gravel and fine sand at CS5 (wide valley section); (b) Size distribution of sediment deposits at CS5, CS13, and CS16; and (c) Boulders, cobbles, and exposed rock bed at CS16 (gorge section).

the low gradient sections. As the uplift further progressed, the low gradient reaches as a result of the lower uplift rate stored more and more gravel and sand. Finally, the sections with low uplift rate stored a massive amount of gravel and sand and formed very wide valleys. This is different from the formation of reservoirs as a result of dam construction, where sediment deposition exhibits clear sorting features: clay and silt deposit near the dams, sand in the middle, and gravel in the upper end of the reservoirs.

1.4 MORPHOLOGICAL CHARACTERISTICS OF RIVERS IN THE TONGDE BASIN

The Tongde Basin is located between the Jishi Gorge and Xining and the Yellow River flows from south to north through the middle of the basin (Fig. 1.1 and 1.21). The Tongde Basin has a drainage area of 3,200 km² and average elevation of 3300 m. It was a graben basin and a huge amount of sediment, consisting mainly of gravel and sand, deposited in the basin. Since 1.2 million years ago the accelerating uplift of the plateau caused retrogressive incision of the Yellow River. About 0.15 million years ago the Yellow River cut through the Tongde Basin (Li et al., 1996). The Tongde basin changed from sedimentation to erosion. Dahaba, Qushian and Baqu rivers incised retrogressively from the confluences with the Yellow River and induced development of new stream networks on the previously sedimentation basin. These stream networks have unique features.

In the late Tertiary period, the lakes in the Tongde, Xining, Guide, Gonghe, and Qinghai Lake basins formed an enormous lake area called the Qingdong Ancient Lakes (Pan, 1994). Later, the uneven uplift and in homogeneity and nonuniform moving velocity of the Qinghai-Tibet Plateau turned the Tongde basin into a fault basin with a large elevation difference between the basin and its surrounding mountains. A large amount of sediment deposits started accumulating within the basin. By the Pliocene epoch to Mid-Pleistocene epoch, a thick layer of lacustrine deposits was evenly distributed within the basin. These deposits mainly consisted of small gravels and pebbles and a large amount of sand. Eolian loess also started accumulating in the basin in the Quaternary period (Zhao, 2005).

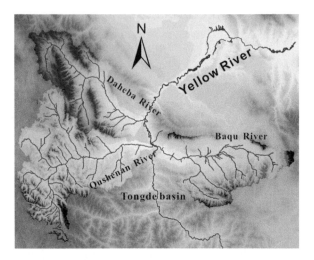

Figure 1.21 River systems developed due to retrogressive erosion in the Tongde Basin.

The intense uplift of the plateau during the Himalaya Orogeny formed the current Yellow River system. The Kunlun-Yellow Orogeny occurred 1.2 million years ago and caused the Yellow River to extend further upstream through headcut erosion. The Yellow River eventually cut through the Jishi Valley. Approximately 50,000 years ago, the Yellow River reached the Xinghai-Tangnaihai area (Liu and Sun, 2007). During the Gonghe Orogeny (0.15 million years ago), the Yellow River incised about 200 m (Yang et al., 1996). The Yellow River cut through the Longyang Velley and entered the Tongde Basin (Li et al., 1996). The incision of the mainstream inevitably resulted in the incision and headcut erosion of the tributaries. The rapid erosion and development of the Qushian, Baqu and Daheba Rivers, tributaries of the Yellow River in the Tongde basin, occurred under this condition.

1.4.1 Development of the Baqu River elongating stream network

The stream network developed in the Tongde Basin is that of an elongating basin as described by Strahler (1964). The Baqu River basin conforms to Valley Type VII, which was characterized by Rosgen (1996, p. 4–15) as follows:

> "Valley Type VII consists of a steep to moderately steep landform, with highly dissected fluvial slopes, high drainage density, and very high sediment supply. Streams are characteristically deeply incised in colluvium and alluvium of in residual soils. … Depositional soils associated with these highly dissected slopes can often be eolian deposits of sand and/or marine sediments."

In the case of the Tongde basin the elongating basin develops in the wide and flat ancient basin consisting of river deposits and eolian loess. The incision in the mainstream causes headcut erosion and the formation of new parallel tributaries which are narrow and short and intercept the deeply-incised mainstream asymmetrically in near right angles. The development of a stream network in an elongating basin requires two basic conditions: 1) there is a wide and flat valley covered with a thick layer of deposits; and 2) a large scale of headcut erosion as a result of deep incisions of the mainstream.

Due to the uneven uplift of the plateau, the Tongde basin became an ancient sedimentary basin. A large amount of river deposits and eolian loess have accumulated in the basin. As a result of the Gonghe Orogeny, the plateau continued to rise, and the ancient Yellow River cut through the Longyang Valley and connected with the other water systems within the Tongde basin. The deposit areas became erosion areas. Rapid incision occurred in the rivers and caused elongate basin stream networks to develop.

A substantial amount of studies have been done in the past regarding the categorization of river systems, classification of river channels, and their development mechanisms (Zernitz, 1932; Horton, 1945; Strahler, 1957). The relations between the river systems and their geomorphological conditions also have been studied (Lubowe, 1964; Small, 1972; Clark et al., 2004). In general, dendritic stream networks primarily develop in areas which are not predominantly impacted by tectonic motion, have relatively consistent lithology and not very inhomogeneous topography, and experience a long term of erosion. And development of a parallel stream network usually indicates that the area has a high gradient and relatively dry climate. The development of a trellis stream network and a roughly rectangular shaped watershed is usually related to geologic folds and faults in the watershed. Stream networks in elongating basins are very different from the traditional river systems in terms of their form, development conditions, and mechanisms. This book discusses this unique river system and provide a new perspective of the morphodynamic processes of the plateau.

1.4.1.1   Development of stream networks in elongating basins due to retrogressive incision

Figure 1.22 shows a typical stream network in an elongating basin and a Rosgen (1996) Valley Type VII in the Yellow River basin upstream from the Longyang Valley. Figure 1.23 shows the Baqu River which is the result of rapid incision. The Baqu River, a tributary of the Yellow River, has experienced extensive headcut erosion due to incisions on the Yellow River. The bank slopes along the downstream reach of the Baqu River have become extremely steep. Water on the overbank areas quickly flows through the original deposit plains and creates many parallel gullies which are narrow on the top and wide at the bottom. From Figure 1.22, it can be seen that the Baqu River channel was originally located above the steep banks of the river valley, but the channel has been extensively incised and the new channel is now tens of meters lower than the original river bed. Rapid incision is the primary driving force for the development of stream networks in elongate basins.

The authors of this section studied the stream networks based on field investigations, on-site measurements, and analyses using remote-sensing images and DEM data. The studies have focused on the locations of the tributary gullies and river valleys, the longitudinal gradients, the lengths of the gullies, and the depth and width of the incision. The development conditions of the stream networks, including the composition and gradation of the river bed materials, were also analyzed. A total of 68 sampling sites were selected, each of which represented one gully in the elongating basin stream

Figure 1.22 (a) Headcut erosion on tributaries due to incisions on the Yellow River; and (b) development of parallel gullies due to incision of tributaries and steepening of bank slopes.

Figure 1.23 Rapid incision of the Baqu River in the Tongde Basin as a result of the uplift of the plateau.

network. All sampling sites were located in the middle to lower reaches of the Baqu River and most of them were located on the first-order streams (gullies). The unique features of the elongating basin stream networks are usually present in the gullies. The first-order gullies usually have an average valley width of only 70 m and length of

Table 1.4 Statistical value of the gullies at 68 sampling sites in the Baqu River.

	Length (m)	Drop in elevation (m)	Connecting angle (°)	Bed gradient (%)	Bank slope (%)
Maximum	1223.5	233.6	132.2	31.5	9.9
Minimum	296.8	29.1	28.1	7.0	4.3
CV (%)	30.5	39.0	29.5	35.9	19.1
Average	766.8	111.4	82.2	15.1	6.1

Note: CV stands for coefficient of variation; Bank slope is the slope of the river overbank area where the gully has developed.

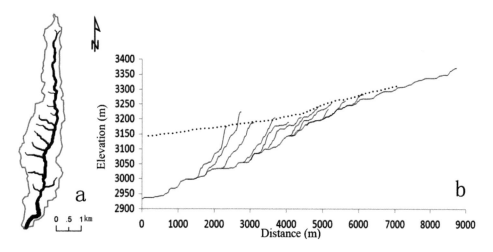

Figure 1.24 (a) Plan view of the stream network in the middle reach of the Baqu River; (b) Longitudinal bed profiles of the Baqu River and parallel gullies (Dashed line is the original deposit basin surface).

less than 1 km (Table 1.4). The gullies are usually smaller than the typical first-order streams of regular rivers. Although the gullies join the mainstream in parallel, the connecting angles are relatively large (Table 1.4). In addition, the parallel gullies occur in the middle to lower reaches of the river mainstream.

Figure 1.24 shows typical plan and profile views of the stream network in the middle reach of the Baqu River. The gullies developed in parallel in the middle to lower reaches of the mainstream with lengths less than 1 km. The longitudinal gradients of the gullies are fairly consistent. The upstream end of all gullies is the original deposit plain of the basin, and, therefore, the gullies are narrow at the upstream end of the mainstream.

Figure 1.25 shows the three elongating stream networks in the middle reach of the Baqu River. The drainage areas of the three watersheds from left to right are 24.9, 17.3, and 7.8 km², respectively. Using the Horton-Strahler stream ordering method, the two larger watersheds are 3rd order, and the smallest one is 2nd order. The Bifurcation Ratio of a typical river network is about 4.5; the Length Ratio is about 2; and the Area Ratio is 3–5 (Craddock et al., 2010; Kinner and Moody, 2005). However, the

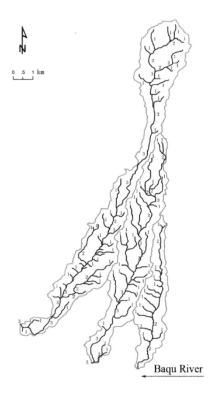

Figure 1.25 Horton-Stahler stream ordering of three elongating basins in the middle reach of the Baqu River.

Bifurcation Ratio of the elongate basins shown in Figure 1.25 is about 10, which is significantly higher than that for general river systems and similar to that of plume networks (Wang et al., 2014). The average Length Ratio and Area Ratio of these elongate basins are also higher than those for general river networks.

1.4.1.2   Development mechanism of elongating stream networks

Howard (1967) stated that the configuration of a river network provides indicators of the tectonic structure and material composition of the area. For instance, the presence of a parallel stream network indicates that there are relatively steep gradients in the area, while a plume stream network typically develops in an area covered with a thick layer of uniform loess. The development of stream networks in elongating basins also is closely related to the environment that they reside in.

The Tongde basin was a deposition basin created by the uneven uplift of the plateau. The topography within the basin is relatively uniform with mild elevation fluctuations. There are many wide valleys inside the basin (Fig. 1.26), which indicates that the overbank areas where the gullies develop have relatively mild gradients. In fact the overbank area was original basin surface and sediment deposition made the surface quite flat. This can be seen from Figure 1.24b. Field measurements also

Figure 1.26 Wide valleys developed on the flat basin surface and comprise elongate stream network.

confirmed that the overbank slopes in the basin have mild gradients, usually no more than 5°, with fairly flat topography and decent vegetation coverage. The gullies also are short and narrow with very small drainage areas (normally less than 0.2 km^2). This vegetation constraint keeps the stream networks from developing further as a result of localized high-intensity rainfall. Therefore, the development of stream networks in elongate basins is most likely a result of headcut erosion, which is different than the development mechanism for other stream network types.

The materials deposited in the Baqu River basin primarily consist of gravels, fine sand, and overlying loess. The sampling particles were relatively small with a median diameter of no more than 10 mm. The gravel particles were relatively round and smooth, which indicated that they had high mobility. In this environment, when scouring occurs under high flows, incision of the river bed will most likely occur. In fact, Yang et al. (1996) and Craddock et al. (2010) presented documents and data from the Tangnaikai and Tongde reaches of the Yellow River and estimated that the mainstream and tributaries of the Yellow River had incised more than 200 m. Pan et al. (2004) stated that the incision of the Yellow River started 50,000 years ago and the rate of incision in the Tongde basin was as high as 400 cm per thousand years.

Another important factor for the formation and stability of stream networks in elongate basins is the erodibility of loess and river deposits which primarily consist of sand and gravels. This composition of the channel materials results in primarily vertical incision instead of lateral erosion under the effects of the river flow, and, therefore, the channel can maintain an overall stability. If the Baqu River basin were covered with erodible soil like the Loess Plateau, or the slope of the overbank areas were steep, or there were a large amount of sand pebbles, the stream network would develop into a more common dendritic network.

1.4.2   Characteristics of the Daheba River

1.4.2.1   The Daheba River and its tributaries

The incision of the Yellow River mainstream has resulted in the rapid incision of the Daheba River, a tributary of the Yellow River (Fig. 1.21). The headcut erosion in the

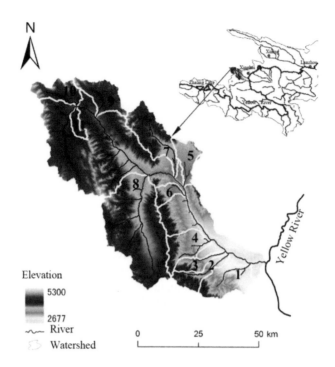

Figure 1.27 Digital elevation of the Daheba River and the locations of the eleven major tributaries.

Daheba River has been gradually extending upstream, which is a primary feature of the development of the Daheba River. Figure 1.27 shows the DEM data for the Daheba basin. The areas in lighter colors are on lower elevations and in the lower reaches of the river. The areas in darker colors are on higher elevations and in the upper reaches of the river. Field investigations were conducted on the Daheba River mainstream and eleven major tributaries. The lower reach of the river is located in a sedimentary basin, while the upper river is located on the mountains. Based on the development environment, the Daheba River was divided into three different zones. They are from lower reach to upper reach in order: the sedimentary basin zone, the transition zone and the mountain zone.

Sedimentary basin zone: During the Pliocene epoch to Mid-Pleistocene epoch, an evenly-distributed and thick layer of river deposits developed in the Tongde Basin. Accumulation of deposits occurred along both banks of the Daheba River. These deposits were several hundred meters thick and primarily consisted of pebbles and sand. The pebbles had good psephicity with relatively consistent sizes (Fig. 1.28). Above a very thick layer of large pebbles mixed with sand, there were layers of coarse and fine sand and loess mixed in the layer of pebbles with sand. It can be seen from the pictures that the river bed consists of pebbles with consistent sizes and good psephicity.

Transition zone: The transition zone is located in the upper reach of the sedimentary basin zone and has been experiencing headcut erosion. This portion of the Daheba River bed has experienced deep incision and cut through the Quaternary deposits, and

Figure 1.28 Overview of the sedimentary basin zone.

eventually entered the bed rock layer. As a result of accelerating incision in the past thousand years, the river channel has become funnel shaped (Fig. 1.29). The top portion is the gradual incision zone and the lower portion is the rapid incision zone. The maximum incision depth within the area of the field investigations was approximately 30 m. The overbank materials in the transition zone consist of three layers: the bottom layer is bed rock; the middle layer is pebbles with sand; and the top layer is loess. In this area, the main bed materials are pebbles of different sizes and collapsed bed rock materials. These bed materials are poorly sorted and have low psephicity. The bed load in this zone differs significantly from the bed materials. The bed load primarily consists of gravel and cobbles and the size of the bed load is between that in the mountain zone and that in the sedimentary basin zone.

Mountain zone: The mountain zone is in the most upper reach of the Badeha River, which is very different from the sedimentary basin zone in the lower reaches. In the mountain zone, bedrock is exposed, and weathered rocks are lying in the river channel. These rocks have irregular shapes with poor psephicity. Some large rocks are lined up by the force of the river flow and develop into step-pool system. The flow through these rock steps is very rapid and then slows down in a deep pool downstream of the steps. Fine sand accumulates in these pools. In general the rivers in the mountain zone have no difference with other mountain rivers.

Jang (1987) stated that under the condition that the earth's crust is under long-term stabilization after a long and homogeneous rate of uplift, the longitudinal profile of the river valley for an idealized drainage network can be described by a power function

$$Y = aX^n,$$

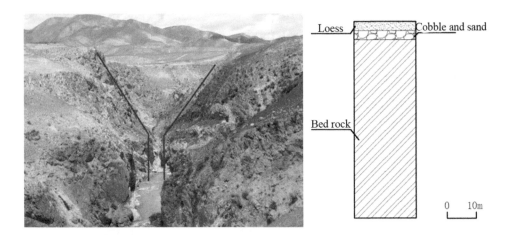

Figure 1.29 (a) Funnel-shaped cross section in the transition zone; (b) Profile of bed materials in the transition zone.

where Y is dimensionless elevation (H/H_0); X is the dimensionless length of the river (L/L_0) from the outlet, and a, n are constants independently determined for each profile. The coefficient n is used as a concavity index. If $n < 1$, the shape of the longitudinal river profile is convex; if $n = 1$, the river profile is a straight line; and if $n > 1$, the shape of the river profile is concave.

The longitudinal profile of the Daheba River is shown in Figure 1.30. The Daheba River has an overall n-value of 0.87. According to Jang (1987), under a typical tectonic condition with steady even uplift of the earth crust, all rivers would experience an erosion cycle. During this cycle, the value of n increases as the river profile changes from a convex parabola during the deep incision period $(n < 1)$ into a near straight line during the transitional period $(n \approx 1)$, and then into a concave parabola $(n > 1)$. Based on this theory, it is apparent that the Daheba River is currently in the deep incision period.

Figure 1.30 shows that the portion of the river profile in the sedimentary basin zone is very different than the portion in the mountain zone. The n-value in the sedimentary basin zone is approximately 1.07, while the value is 1.60 in the mountain zone and 0.79 in the transition zone. This means that the portion of the river in the sedimentary basin zone has already gone through the deep incision period and is now moving into a stable period; the mountain zone is in the stable period that developed before the incision of the Yellow River started; and the transition zone is currently in the deep incision and headcut period.

Computation of the n-value of 11 tributaries of the Daheba River basin showed that the development of the tributaries was consistent with the development of the mainstream. For instance, tributary Nos. 1, 2, and 3 were located in the sedimentary basin zone. The n-value of these three tributaries is 1.25, 1.20, and 1.04, respectively. All of these values were around 1, meaning that these tributaries were transitioning into a stable period. Tributary Nos. 9, 10, and 11 are located in the mountain zone with n-value of 1.37, 1.41, and 2.69, respectively. The n-values of these tributaries are

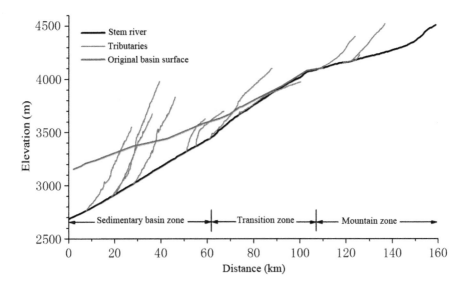

Figure 1.30 Longitudinal profile of the Daheba River and its tributaries.

significantly higher than those of the tributaries in the sedimentary basin zone and are all higher than 1, meaning that these tributaries are in a relatively stable period.

1.4.2.2   Development of debris flow gullies

The primary contents of the deposits in the sedimentary basin zone are sand and pebbles with good psephicity. Due to the lack of loess, there is poor vegetation cover. Debris flow occurs frequently in the zone due to erosion caused by rain storms. At the same time, the Daheba River is experiencing erosion incision, which expedites debris flows. The catchment area of the river expands following debris flow gullies developing onto overbank flat original basin surface. Field investigations found that there were approximately 60 debris flow gullies along the left bank of the Daheba River. A typical debris flow gully is shown in Figure 1.31a, which flows into the Daheba River. The upper end of the debris flow gully cut into the original basin deposit and form a sharply incised ditch (Figure 1.31b). The gradients of these debris flow gullies are quite uniform, mostly between 8–12%.

Figure 1.32 shows the bed profiles of 9 debris flow gullies on the left bank of the Dabeha River and a plan view of a typical debris flow gully basin. The bed gradients of these gullies are almost the same because the sediment is the same in all the debris flow gullies. There is a sharp change at the original basin surface where the gradient becomes much smaller and is nearly equal to that of the basin surface. The lengths of these gullies are short, with longest length only 4 km near the river mouth. The debris flow gullies with distance from the mouth are shorter and narrower. The debris flow gullies are mostly second order stream in Hoton's stream ordering system and have a few tributaries, which are about 500 m in length. The sediment particles of the debris flow deposit are quite uniform and similar to that of the bed load in the river. In fact

Figure 1.31   (a) A debris flow gully flows into the Daheba River; (b) Upper end of a debris flow gully.

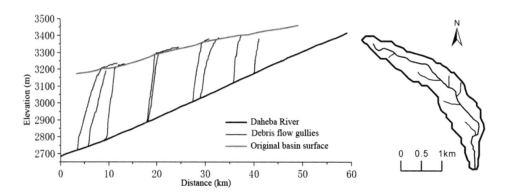

Figure 1.32   (a) Bed profiles of 9 debris flow gullies on the left bank of the Dabeha River; (b) Plan view of a typical debris flow gully basin.

these debris flow gullies are source of bed load for the Daheba River. A lot of sediment from the debris flow gullies is transported into the Daheba River, which has caused the river to develop into a braided river with a wide river valley and a large number of gravel bars.

Although both the Deheba and Baqu rivers are located within the Tongde basin, debris flow gullies developed along the Daheba River, while elongating basin streams developed along the Baqu River. The main reason is the differences in compositions, layer thickness, and vegetation. The Baqu River basin has relatively good vegetation. As a result, narrow and short gullies develop and form elongate basin stream networks. In the Daheba River watershed, pebbles with good psephicity make up the majority of the deposit with a fairly thin loess layer. The vegetation is poor. Therefore, debris flow occurs due to serious erosion during high floods.

REFERENCES

Burbank, D.W., Leland, J. and Fielding, E. 1996. Bedrock incision, rock uplift and threshold hillslopes in the northwestern Himalayas. Nature, 379: 505–510.

Chen, G.D. 1997. Mechanism and Historical Background of the uplift of the Qing-Xizang(Tibet) Plateau. Geotectonica et Metallogenia, 21(2):95–108 (in Chinese).

Chung, S.L., Lo, C.H., Lee, T.Y., Zhang, Y., Xie, Y., Li, X., Wang, K.L. and Wang, P.L. 1998. Diachronous uplift of the Tibetan plateau starting 40M years ago. Nature, 394: 769–774.

Clark, M.K., Schoenbohm, L.M., Royden, L.H., Whipple, K.X., Burchfiel, B.C., Zhang, X., Tang, W., Wang, E. and Chen, L. 2004. Surface uplift, tectonics, and erosion of Eastern Tibet from large-scale drainage patterns. Tectonics, 23(1): TC1006, doi: 10.1029/2002TC001402.

Coleman, M. and Hodges, K. 1995. Evidence for Tibetan Plateau uplift before 14 Myr ago from a new minimum age for east–west extension. Nature, 3742: 49–52.

Craddock, W.H., Kirby, E., Harkins, N.W., Zhang, H.P., Shi, X.H. and Liu, J.H. 2010. Rapid fluvial incision along the Yellow River during headward basin integration. Nature Geosciences, 3: 209–213.

Cui, Z.J., Gao, Q.Z., Liu, G.N., Pan, B.T. and Chen, H.L. 1996. The Planation Surface, Palaeokarst and the Uplift of Qinghai-Tibet Plateau. Science in China (Series D), 26(4):378–386 (in Chinese)

Dodds, P.S. and Rothman, D.H. 2000. Scaling, universality, and geomorphology. Annu. Rev. Earth. Planet. Sci. 28:571–610.

Dong, X.B., Wang, Z.M., Tan, C.Z., Yang, H.X., Cheng, L.R. and Zhou, Y.X. 1991. New results of paleomagnetic studies of the Qinghai-Tibet P lateau. Geological Review, 37(2):160–164 (in Chinese).

Gao, R., Xiong, X.S., Li, Q.S. and Lu, Z.W. 2009. The Meho Depth of Qinghai-Tibet Plateau Revealed by Seismic Detection. Acta Geoscientica Sinica, 30(6):761–773 (in Chinese).

Gupta, S. 1997. Himalayan drainage patterns and the origin of fluvial mega fans in the Ganges foreland basin. Geology, 25(1): 11–14.

Hallet, B. and Molnar, P. 2001. Distorted drainage basins as markers of crustal strain east of the Himalaya. J Geophys Res, 106(13): 697–709.

Han, Q.W. 2009. Analysis on bed load amount and research on sedimentation of Three Gorges Reservoir. Water Resources and Hydropower Engineering, 40(8): 44–55. (in Chinese).

Harrison, T.M., Copeland, P., Kidd, W.S.F. and Yin, A. 1992. Raising Tibet. Science, 255: 1663–1670.

Harvey, A.M. and Wells, S.G. 1987. Response of Quaternary fluvial systems to differential epeirogenic uplift: Aguas and Feos river systems, southeast Spain. Geology, 15: 689–693.

Horton, R.E. 1945. Erosional development of streams and their drainage basins: Hydrophysical approach to quantitative morphology. Geological Society of America Bulletin, 56(3): 275–370.

Howard, A.D. 1967. Drainage analysis in geologic interpretation: A summation. The American Association of Petroleum Geologists Bulletin, 51(11): 2246–2259.

Jang, Z.X. (1987). Model of development and rule of evolution of the longitudinal profiles of the valley of three rivers in the northwestern part of Yunnan province. Journal of Geographical Sciences, 42(1), 16–27 (in Chinese).

Kinner, D.A. and Moody, J.A. 2005. Drainage networks after wildfire. International Journal of Sediment Research, 20(3): 194–201.

Lavé, J. and Avouac, J.P. 2001. Fluvial incision and tectonic uplift across the Himalayas of central Nepal. J Geophys Res, 106(B11): 26561–26591.

Li, J.J. 2004. On the uplift of Tibetan Plateau. Proceedings of Danxiashan conference, Quaternery Committee, Chinese Association of Geography, China Environmental Science Press, Beijing, 1–4 (in Chinese).

Li Siguang, 1999. Introduction to geological mechanics, Geology Press, p. 133, Beijing (in Chinese).

Li, T., Chen, B.W. and Dai, W. 2010. Geological atalas of Qinghai-Tibet Plateau and neighboring areas in different uplifting stages. Guangdong Science Press, Guangzhou (in Chinese).

Li, Y.L., Wang, C.S., Wang, M., Yi, H. and Li, Y. 2006. Morphological features of river valleys in the source region of the Yangtze River, northern Tibet, and their response to neotectonic movement. Geology in China, 33(2): 374–382. (in Chinese).

Li Yandong, 1995. The process and mechanism of uplift Qinghai-Tibet plateau. Acta Geoscientica Sinica, 1995(1):1–9 (in Chinese).

Li, J.J., Fang, X.M., Ma, H.Z., et al., 1996. The evoltion of the Yellow river geomorgraphy and the uplift of Tibetan plateau. Science in China, Ser. D, 26(4): 316–322 (in Chinese).

Li, J.J. and Fang, X.M., Uplift of the Tibetan Plateau and environmental changes, Chinese Science Bulletin, 1999, 44(23): 2117–2124 (in Chinese).

Li, J.J., S.X. Wen, Q.S. Zhang, F.B. Wang, B.X. Zheng, B.Y. Li, A discussion on the period, amplitude and type of the uplift of the Qinghai-Xizang Plateau, *Scientia Sinica*, 22, 1314–1328, 1979 (in Chinese).

Li, J.J. 1983. Morphological framework and genetie mechanism of the Qinghai-Xizang Plateau. Mountain Reaserch, 1(1):7–15 (in Chinese).

Li, Y., Densmore, A.L., Zhou, R.J., Ellis, M.A., Zhang, Y. and Li, B. 2006. Profiles of digital elevation models (DEM) crossing the eastern margin of the Tibetan Plateau and their constraints on dissection depths and incision rates of late cenozoic rivers. Quaternary Sciences, 26(2):236–243 (in Chinese).

Liu, H.X. and Wang, Z.Y. 2007, Morphological feature and distribution of typical river networks. ShuiLi XueBao, 34(11):1354–1357 (in Chinese).

Liu, H.X. and Wang, Z.Y. 2008, Relationship between river network pattern and enviromental condition. J Tsinghua Univ (Sci & Tech),48(9):29–32 (in Chinese).

Liu, Z.J. and Sun, Y.J. 2007, Uplift of the Qinghai-Tibet Plateau and Formation, Evolution of the Yellow River. Geography and Geo-Information Science, 23(1):79–91 (in Chinese).

Liu, Z.X., Su, Z., Yao, T.D. Wang, W.D. and Shao, W.Z. 2000, Resource and Distribution of glaciers on the Tibetan Plateau. Resource Science, 22(5):49–52 (in Chinese).

Lubowe, J.K. 1964. Stream junction angles in the dendritic drainage pattern. American Journal of Science, 262: 325–339.

Ming, Q.Z. 2006, The Landform Development and Enviroment Effects in the area of Three Paralle Rivers, Northern Longtitudinal Range-Gorge Region(Lrgr). Lanzhou University (in Chinese).

Mo, X.X. 2011, Magmatism and Evolution of the Tibetan Plateau. Geological Journal of China Universities, 17(3):351–367 (in Chinese).

Pan, B.T. 1994, A study on the Geomorphic Evolution and Development of the upper reaches of Yellow River in Guide Basin. Arid Land Geography, 17(3):43–50 (in Chinese).

Pan, B.T., Gao, H.S., Li, B.Y. and Li, J.J. 2004, Step-Like landforms and Uplift of the Qinghai-Xizang Plateau. Quaternary Sciences, 24(1):50–57 (in Chinese).

Qin, D.H., Yao, T.D., Zhou, S.Z., Feng, Z.D., Chen, F.H., Fang, X.M., Pan, B.T. and Wang, N.Y. 2013. Journal of Lanzhou University (Natural Sciences), 49(2):147–153 (in Chinese).

Rea, D.K. 1992. Delivery of Himalayan sediment to the Northern Indian Ocean and its relation to global climate, sea level uplift and seawater strontium. Geophysics Monograph, 70: 387–402.

Rosgen, D. 1996. Applied River Morphology. Wildland Hydrology, Pagosa Springs, Colo., USA, variable pagination.

Rowley, D.B., Currie, B.S. 2006. Palaeo-altmietry of the Late Eocene to Miocene Lunpola basin, Central Tibet. Nature, 439: 677–681.

Royden, L.H., Burchfiel, B.C., van der Hilst, R.D. 2008. The geological evolution of the Tibetan Plateau. Science, 321: 1054–1058.

Small, R.J. 1972. The study of landforms. Cambridge University Press, Cambridge, 486p.

Smart, J.S. 1972. Quantitative characterization of channel network structure. Water Resources Research, 8(6): 1487–1496.

Spicer, R.A., Harris, N.B.W., Widdowson, M.W., Herman, A.B., Guo, S.X., Valdes, P.J., Wolfe, J.A., Kelley, S.P. 2003. Constant elevation of Sourthern Tibet over the past 15 million years. Nature, 421: 622–624.

SSBAF (State Seismology Bureau Artun Active Faults Team), 1992. Artun active faults, Seismology Press, Beijing (in Chinese).

Strahler, A.N. 1957. Quantitative analysis of watershed geomorphology. Transactions of American Geophysics Union, 33: 913–920.

Strahler, A.N. 1964. Part II. Quantitative geomorphology of drainage basins and channel networks, Section 4-II in Chow, V.T. ed., Handbook of Applied Hydrology, McGraw-Hill, New York, variable pagination.

Tapponnier, P., Xu, Z.Q., Roger, F., Meyer, B., Arnaud, N., Wittlinger, G. and Jingsui, Y. 2001. Oblique stepwise rise and growth of the Tibet Plateau. Science, 294(5547): 1671–1677.

TECAS (Tibetan Expedition team of Chinese Academy of Sciences), 1983. Tibetan geomorphology, Beijing: Science Press. (in Chinese).

TECAS (Tibetan Expedition team of Chinese Academy of Sciences), 1984. Rivers and lakes in Tibet, Beijing: Science Press. (in Chinese).

Teng, J.W. Wang, S.Z. Yao, Z.X., Xu, Z.W., Zhu, Z.W., Yang, B.P. and Zhou, W.H. 1980, Charateristics of the Geophysical Fields and Plate tections of the Qinghai-Xizang Plateau and Its neighbouring regions. Acta Geophysica Sinica, 23(3):254–267 (in Chinese).

TSIT-CAS (Tibet Plateau Comprehensive Scientific Investigation Team of Chinese Academy of Sciences), 1983. Geomorphology of Tibet, Chinese Science Press, Beijing (in Chinese).

Tumer, S., Hawkesworth, C., Liu, J., Rogers, N., Kelley, S., van Calsteren, P. 1993. Timing of Tibetan uplift constrained by analysis of volcanic rocks. Nature, 364: 50–54.

Walcott, R.C. and Michael, A. 2009. Universality and variability in basin outlet spacing: Implications for the two-dimensional form of drainage basins. Summerfield Basin Research, 21: 147–155.

Wang, Z.Y., Li, Y.T. and He, Y.P. 2007. Sediment budget of the Yangtze River. Water Resources Research, 43: 1–14.

Wang, Z.Y. and Zhang, K. 2012. Principle of equivalency of bed structures and bed load motion. Int J Sediment Res, 27(3): 288–305.

Wang, Zhao-Yin, Lee, J.H.W. and Melching, S. 2014. River dynamics and integrated river management, Springer Verlag and Tsinghua Press, Berlin/Beijing. pp. 1-800 ISBN 978-3-642-25651-6.

Wang Zhaoyin, Yu Guoan, Wang Xuzhao, Charles S. Melching and Liu Le, 2015. Sediment storage and morphology of the YaluTsangpo valley due to uneven uplift of the Himalaya.Science China Earth Science, 58(8): 1440–1446.

Wang, Z.Y. and his team, 2015. Field investigation of the Yarlung Tsangpo River. State Key Lab of Hydroscience and Engineering, Tsinghua University (in Chinese).

Xiao, X.C. and Wang, J. 1998, A Brief Review of Teetonic Evolution and Uplift of the Qinghai–Tibet Plateau.Geological Rivew, 44(4):372–381 (in Chinese).

Xu, R. Tao, J.R. and Sun, X.Q. 1973. On the discovery of a quercus semicarpifolia bed in mount shisha pangma and its significance in botany and geology. Journal of Integrative Plant Biology, 15(1):103–119 (in Chinese).

Yang, D.Y., Wu, S.G. and Wang, Y.F. 1996, On River Terraces of the upper reaches of the Huanghe River and Change of the River System. Scientia geographica Sinica, 16(2): 137–143 (in Chinese).

Yin, A. and Harrison, T.M., 2000. Geological evolution of Himalayan-Tibetan Orogen, Ann. Rev. Earth Planet Sci., (28) 211–280.

Zernitz, E.R. 1932. Drainage patterns and their significance. Journal of Geology, 40: 498–521.

Zhang, D.D. 1998. Geomorphological problems of the middle reaches of the Tsangpo river, Tibet. Earth Surf Process Landf, 23: 889–903.

Zhang, P.Z., Shen, Z.K., Wang, M., Gan, W.J., Bürgmann, R., Molnar, P., Wang, Q., Niu, Z.J., Sun, J.Z., Wu, J.C., Sun, H.R. and You, X.Z. 2004. Continuous deformation of the Tibetan Plateau from global positioning system data, Geology, 32: 809–812.

Zhao, Z.M. and Liu, B.C. 2005. The primary perspective of Longyang Gorge formation. Northwestern Geology, 38(2):24–13 (in Chinese).

Zhong, D.B. and Ding, L. 1996. A discussion on the process and mechanism of uplift of Qinghai-Tibet plateau. Science In China(Series D), 26(4):289–295 (in Chinese)

Chapter 2

Fluvial processes of incised rivers in Tibet

The Himalaya Mountains and Qinghai-Tibet Plateau have been rising for 3.4 million years and the rising rate has been increasing since late Cenozoic (Coleman and Hodges, 1995; Burbank et al., 1996; Chung et al., 1998; Yin, 2000; Tapponnier et al., 2001; Royden et al., 2008). Continuous and ever-increasing uplift of the plateau accelerates fluvial incision and rates of mass movement of the rivers on the plateau, and increases sediment transport by the rivers (Harvey and Wells, 1987; Lavé and Avouac, 2001).

Most rivers (or river reaches) in southeast Tibet, especially those in the lower Yarlung Tsangpo basin on the edge of the plateau, are deeply incised. The deep incision of river channels, along with high regional annual precipitation due to the influence of Indian monsoon and the fragmented landform due to strong neo-tectonic movements, causes frequent occurrence of avalanches, landslides, and debris flows in these region and inevitably exerts significant influence on the fluvial processes of the rivers. Moreover, there are several thousands of glaciers in the plateau, which are mostly very active. The glacial processes affect the fluvial processes and change the river morphology.

This chapter examines the fluvial processes of incised reaches of the Yigong and Palong Tsangpo rivers, the two largest and most important tributaries in the lower Yarlung Tsangpo basin. The discussions in this chapter focus on the effects of natural events (avalanche, landslide, debris flow, and glacial erosion) on river morphology, including sedimentation or incision, development of channel planform pattern, change of longitudinal and cross sectional profiles. The analyses have been conducted based on typical case studies, i.e., the Yigong Tsangpo River (landslide), the Guxiang and Tianmo Gullies (debris flows) and the Palong Tsangpo River (glacial processes).

2.1 LANDSLIDES AND DEBRIS FLOWS

2.1.1 Spatial distribution of landslides and debris flows

Under the combined influence of the regional geology, geomorphology and climate features, numerous rock avalanches, landslides, and debris flows develop in the lower Yarlung Tsangpo basin, which mainly distribute along the Palong and Yigong Tsangpo rivers and the Yarlung Tsangpo Grand Canyon (Fig. 2.1).

Avalanches often occur along the incised reaches of the Yarlung Tsangpo River and its tributaries (Palong and Yigong Tsangpo rivers) (Fig. 2.2). Compared to landslides and debris flows, the volume of avalanches is generally much smaller. Materials from the avalanches occasionally dam the river channels and form dammed lakes, however,

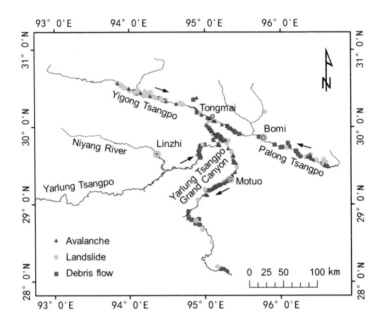

Figure 2.1 Distribution of avalanches, landslides, and debris flows in the lower Yarlung Tsangpo basin.

Figure 2.2 Avalanches often occur along highly incised reaches of the Yarlung Tsangpo River and its main tributaries. (a) Avalanche in the Grand Canyon of Yarlung Tsangpo; and (b) Avalanche in the lower Palong Tsangpo River.

the lakes are often small and the dams are easily breached. Quite often, large-sized materials (big rocks and boulders) fall into the river as result of avalanches and develop (form) into streambed structures, which are beneficial for the stability of the channel bed and bank slopes.

2.1.2 Effects of the Yigong landslides on fluvial processes

2.1.2.1 Two landslides

The fluvial process of the Yigong Tsangpo River (Fig. 2.3) has been influenced by two super-large landslides in the last two centuries, one occurred in 1900 and the other in

Table 2.1 Key features of the two landslides in the Yigong Tsangpo River.

Landslides	Occurrence time	Outbreak time	Landslide volume ($10^9 m^3$)	Dam height (m)	Dam width (m)	Dam length (m)
1900	Summer, specific time not clear	One month later	0.51	75–140	1,200–2,500	3,200–3,600
2000	8:00 pm, April 9	June 10	0.30	60–100	2,200–2,500	1,000

Data sources: Yin, 2000; Shang et al., 2003.

2000. The landslides carried a massive amount of materials into the Yigong Tsangpo River. After years of continuous scouring and sorting, particles of big sizes (boulders and cobbles) gradually formed step structures, which restrain further incision of the river channel.

Several literatures have reported the two landslides and analyzed their causes and effects (c.f., Yin, 2000; Zhou et al., 2001; Shang et al., 2003; Evans and Delaney, 2011). Both Yigong landslides occurred almost at the same site which is located in the downstream reach of the Yigong Tsangpo River (Fig. 2.3). The landslide in 1900 caused a distal debris flow which rushed through and reached the opposite slope of the Yigong Tsangpo valley, forming a sediment fan and a landslide dam (Shang et al., 2003). The landslide in 2000 occurred in the Zhamunong Gully (Figs. 2.3, 2.4). The position of the landslide center was 30°12′3″N, 94°58′03″E. The landslide lasted about 10 minutes and formed a massive landslide body with a total length of about 8,000 m and a total volume of about 0.3×10^9 m^3. The elevation of the upper end edge of the landslide was 5520 m, while the channel bed was at about 2190 m (hence the elevation difference of the entire landslide was about 3330 m). Table 2.1 shows the basic data of the two landslides.

2.1.2.2 *Fluvial processes after the landslides*

The ETM+ remote sensing images in 1991, 1996, 2000 (7 months after the landslide), 2006, 2009, and 2014 were collected to show the planform change of the channel. Field investigations were conducted to examine the channel pattern and sediment composition. Eleven cross sectional profiles (CS 1 to CS 11, Figs. 2.4 and 2.5) were selected along the river from upstream to downstream to compare the planform pattern changes. ASTER GDEM data and Google Earth images were combined to extract the longitudinal profile of the channel (Fig. 2.5). Table 2.2 shows the basic features of these cross sections.

CS 1 is located most upstream from the dammed lake. At this location, the river is a single-thread channel and free of influence of landslide effects. CS 2 to CS 5 are located in the landslide dammed lake. The river channel near these sections is unstable and wandering, especially at the locations near CS 3, 4 and 5 (Fig. 2.6 (a)). The channel braiding index (defined here as the number of channels at a cross section) first decreased in 2000 after the landslide and then increased in 2006, 2009 and 2014. CS 6

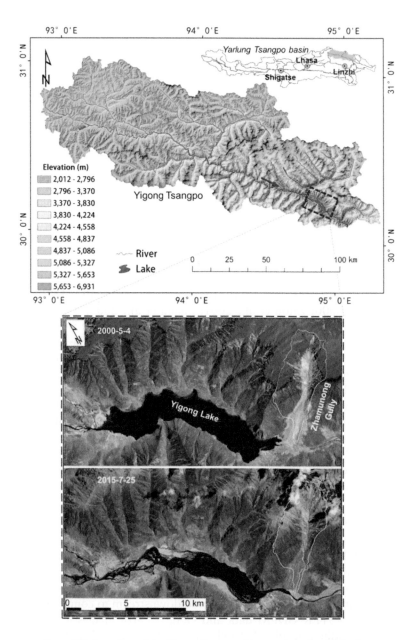

Figure 2.3   Location of the Yigong Tsangpo River and the landslide site.

is located at the landslide dam, and CS 7 to CS 11 are located downstream of the dam. The braiding index of the channel between CS 7 and 11 basically had no change. The channel near CS 6 to CS 8 is a single-thread (Fig. 2.6 (b)). It was silted up by landslide materials and then gradually incised in recent years and the planform has had little change (Fig. 2.4).

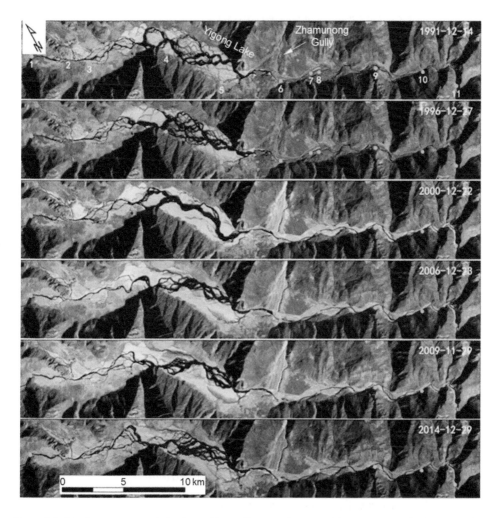

Figure 2.4 Planform pattern of the lower Yigong Tsangpo River in different periods before and after the landslide occurred in 2000.

The longitudinal profile and planform patterns show significant difference along the river reach before and after the landslide. The reach from CS 1 to CS 5 (upstream of the landslide dam, knickpoint) basically changed from a single-thread channel to a braided channel. The reach from CS 2 to CS 5 (located in the dammed lake) formed a wide valley with shallow water depth. At downstream of the landslide dam (knickpoint), the channel was still confined as before even though a huge flood occurred due to the break of the landslide dam, which crashed the valley banks to some extent and increased the width of the channel.

Figure 2.7 shows the sediment composition at cross sections CS 1, 3, 5 and 7. The channel sediment composition spatially changes upstream to downstream from mainly boulders (CS 1), cobbles and gravel (CS3), silt and clay (CS5), to boulders (CS 7) again.

Table 2.2 Basic features of selected cross sections.

CS No.	Width of valley bottom (m)	D50 (mm)	Local gradient (m/m)
1	166	160	0.0075
2	267	135	0.0069
3	3157	30	0.0063
4	2,450	0.068	0.0023
5	2,100	0.068	0.0018
6	696	21	0.0158
7	702	410	0.0158
8	676	680	0.0147
9	398	175	0.0026
10	276	500	0.0075
11	296	330	0.0077

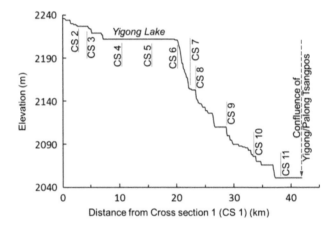

Figure 2.5 Longitudinal profile of the lower Yigong Tsangpo River.

Detailed sediment size distributions at the chosen cross sections are shown in Fig. 2.8. Cross sections CS 4 and 5 are located in the dammed lake. The channel at this location is composed of silt, fine sand and clay and is much finer than the sediment composition at other sections. Cross sections CS 2 and 3 are located upstream of the lake. The size of the channel bed sediment at this location is partly influenced by the dammed lake. The bed material at CS 1 is composed of boulders, cobbles and gravels. The bed material at CS 2 and 3 is composed of cobbles, gravels, and sand. Cross section CS 6 is located at the landslide dam site. At this location, the material is composed of particles with sizes from as big as large stones (1 ∼ 5 m) to as small as clay (<0.01 mm). Downstream of the landslide dam site (CS 6), the size of the deposited material is quite large (CS 7, 8 and 9) but gradually becomes finer due to flow sorting effects which has scoured and transported small particles of the landslide materials to the downstream reach.

Figure 2.6 Different planform patterns at upstream (a) and downstream (b) of the landslide dam (a) Braided channel near CS 5; and (b) Single-thread meandering near CS 8.

Figure 2.7 Channel bed composition at different locations of the Yigong Lake (a) near CS 1; (b) near CS 3; (c) near CS 5; and (d) near CS 7.

2.1.2.3 Sedimentation processes

When the new landslide dam formed in 2000, the flow velocity upstream of the dam reduced and caused further sediment deposition and siltation of the channel bed. Bore coring samples of the sediment deposits from the dammed lake were taken during a field survey in April 2014 (Fig. 2.9). Results from the survey showed that the deposition depth caused by the landslide in 2000 was roughly between 1.25–2.85 m, and the

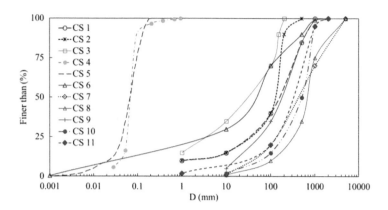

Figure 2.8 Size distributions of channel bed materials at different cross sections upstream and downstream of the Yigong Lake.

Figure 2.9 Bore coring probe profile in the Yigong dammed lake.

average annual siltation rate since 2000 was about 0.1 m. Another interesting finding was that there was a gravel bed belt intruding from the upstream end of the dammed lake (near CS 1) to downstream, and beneath the gravel belt was a deep layer of sand deposition. This may indicate that the deposition of the Yigong Lake has already reached its equilibrium since the occurrence of the landslide in 2000.

Downstream of the landslide dam, two distinct terraces were found for a length of about 10 km along the river. The solid lines (white and red) in Fig. 2.10 (a) and (b) show the deposition plains which formed after the landslide in 1900, while the dashed lines (yellow and blue) in the figures show the deposition plains formed after the landslide in 2000 (Fig. 2.10 (c) and (d)). The break of the landslide dam in 2000 transported a massive amount of sediment to the downstream reach, and formed a very large siltation body. The sediment siltation body has been continuously scoured

Figure 2.10 Deposition plains downstream of the landslide dams.

and incised by as much as 11–13 m since 2000 with a mean incision rate of 1 m/a (Fig. 2.10 (c) and (d)).

The breach of the Yigong landslide dam in 2000 resulted in a massive flood which caused very serious scouring and destroying of two bank slopes downstream of the dam. Fig. 2.11 shows the obvious scouring belt caused by the flood. The belt has an elevation about 20 m higher than the present river bed.

2.1.3   Debris flow dams

In southeastern Tibet, debris flows repeatedly block rivers and form dammed lakes, such as the debris flow occurred in the Guxiang Gully in 1953, the debris flow in the Peilong Gully in 1984, and two debris flows in the Suotong Gully from 2007 to 2009 and recently in May 2011. The debris flows occurred in the Tianmo Gully in recent years also blocked the Palong Tsangpu River and formed a small dammed lake (Fig. 2.12).

The Guxiang Gully (the gully tongue is located at E95°27′06″, N29°54′25″) is a right-bank tributary of the Palong Tsangpo River. With a watershed area of 25.2 km², the gully is surrounded by glacial alpine and has a main channel length of 8.7 km and an average gradient of 25.6%. Continuous high temperature in 1953 led to a large

Figure 2.11 Scouring belt and remnants (red line) along the river downstream of the Yigong landslide dam after the massive flood caused by the breach of the dam in June 2000.

Figure 2.12 Location map of the Guxiang and Tianmo gullies and the Guxiang Lake.

glacial debris flow, which blocked the Palong Tsangpo River and formed the Guxiang Lake. The Guxiang Lake reduced the flow gradient of the Palong Tsangpo River. A massive amounts of sediment deposited in the lake, controlled the channel incision and changed the channel planform from single-thread to braided. Sediment deposition rate in the lake was studied to reveal the channel pattern evolution features in this debris flow dammed lake.

Sediment deposition samples were taken from the Guxiang Lake delta (drilling site shown in Fig. 2.12). The drilling was extended to a depth of 246 cm below the bed surface. Deeper depth was not detected due to the influence of groundwater. The particle sizes (every 5 mm long) of the core samples were analyzed. In general, the particle size distribution can be considered to be the same as the distribution of the water flow environment, hence it was marked using the same color boxes in Fig. 2.13. The black sections of the sand cores are shallow and represent abundant

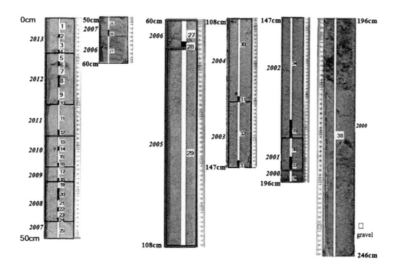

Figure 2.13 Sediment deposition thickness of the Guxiang debris-flow dammed lake.

water periods while the white sections of sand cores are deep and represent low water periods.

The deposited sediment in the Guxiang Lake is mostly less than 1 mm in diameter with fine sand being the dominant component. Sedimentation layers with coarse sand, fine sand and clay content interbed and often develop into uniform and shallow horizontal bedding. In the wet season, the deposition is dominated by fine sand layers which are in light colors and large thicknesses. In the dry season, the deposition is primarily composed of silty clay layers which are in dark colors and small thicknesses. The annual sedimentary rhythm is obvious and colors and structures of the sediment layers are alternant. According to related research on modern sedimentary rhythms types (Glenn and Kelts, 1991), the sedimentation layer in the Guxiang Lake is classified as the fine sand – silty clay layer type. The core samples were divided into different years, based on the colors of the deposit and the particle size distribution (Fig. 2.13). Some of the layers were thick, which were related to special hazard events. For example, the sediment deposition layer for 2005 was quite thick. The main reason was that large scale debris flows broke out from the Guxiang Gully in 2005, which caused marked increase of the height of the debris flow dam and consequently serious sediment deposition in the lake.

The local precipitation has a significant effect on the sediment deposition. Precipitation strengthens the erosion intensity and increases the runoff, which is beneficial for transporting coarse particle materials into the lake. Therefore, during flood periods, coarse sediment is transported into the lake; during the dry season, only fine sediment is transported into the lake; and during normal water period, the deposition of sediment sizes ranges between those in the flood and dry seasons (Fig. 2.14).

Upstream of the knickpoint (the debris flow dam), wide and shallow river channel formed since huge amount of sediment is deposited behind dam. The channel pattern is mainly braided and prone to be wandering. While downstream of the knickpoint, the

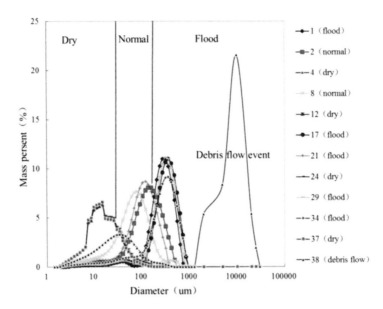

Figure 2.14 Particle size distribution of sand samples drilled in the Guxiang Lake.

Figure 2.15 Sediment composition upstream and downstream of the Guxiang Lake (a) Fine particle deposits upstream of the Guxiang debris flow dam; and (b) Torrents downstream of the debris flow dam (knickpoint). Channel bed is composed of boulders and large cobbles (the young man with height of 175 cm).

river channel of the Palong Tsangpo is deeply incised, and the channel bed is mainly composed of boulders and cobbles (Fig. 2.15).

Another typical case of debris flow's influence on the river morphology is the debris flows from the Tianmo Gully, a left-bank tributary of the Palong Tsangpo River which joins the Palong Tsangpo River at a location about 18.5 km downstream of the Guxiang derbis flow dam (Fig. 2.12). It was reported that a recent large-scale debris flow in the Tianmo Gully occurred in September 2007. The debris flow formed

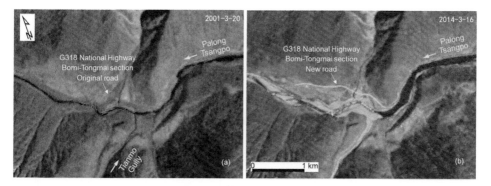

Figure 2.16 Tianmo Gully and the river channel (a) before and (b) after the debris flows.

Figure 2.17   Debris flow from the Tianmo Gully pushed the river flow to the right bank of the Palong Tsangpo River and caused serious erosion and collapsing of the right bank.

a dam across the Palong Tsangpo River and pushed the river flow to the opposite side of the river channel (right bank), which caused serious scouring and collapsing of the right bank. The original road of the G318 National Highway (Bomi – Tongmai section) was destroyed and had to be changed to higher elevations due to this event (Figs. 2.16 and 2.17). The height of the debris flow dam was estimated to be 45 m in 2011 at a field investigation, however, a later field investigation in 2013 found it to have decreased to 38 m due to continuous scouring.

2.2 GLACIAL PROCESSES

The southeast edge of the plateau is a region where numerous oceanic glaciers have developed. The environment in this region has obviously been affected by the following features: (1) it is located at the outlet of the "water vapor channel" formed by the Yarlung Tsangpo Grand Canyon. Plenty of precipitation occurs annually in this region since numerous water vapor from the Indian Ocean arrives here, and the glaciers in this region are experiencing high movement rates; and (2) the Nangabawa, one of the two super knotty structures in this area, is one of the regions on the plateau with

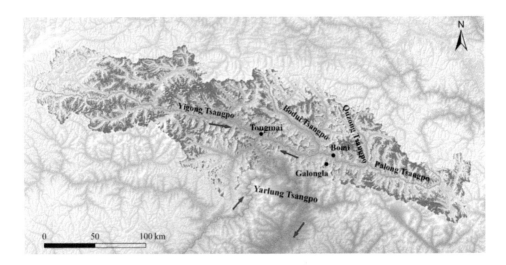

Figure 2.18 Glacier distribution in the Palong Tsangpo basin (Remote sensing data: ETM+).

the highest uplift rate and the largest relative elevation difference. Hence, this region has high elevation, steep relief, high rock stress, and is easy to be eroded. The glacial movement transports a huge amount of materials into the gullies and rivers, in addition, the sediment concentration of the glaciers in this region is much higher than that in other regions due to its unique climate and geologic and geomorphologic environment.

2.2.1   Glaciers in the Palong Tsangpo basin

2.2.1.1   Distribution of glaciers in the Palong Tsangpo

The Palong Tsangpo basin (including the Palong and Yigong Tsangpos) is the largest and most important tributary basin in the downstream area of the Yarlung Tsangpo River basin. Its main stem river has a length of 270 km with a mean longitudinal gradient of 1.19%. Numerous glaciers distribute along the Palong and Yigong Tsangpos (Fig. 2.18), and the area of a single glacier is usually small (about several to a dozen km^2). The material equilibrium line of the glaciers is often at a relatively low elevation, and prone to climate change. A very significant amount of moraine deposits distributes in the basin, especially in the gullies along the main tributaries.

The longitudinal profile and gradient along the Palong Tsangpo River are shown in Fig. 2.19. The river can be divided into 4 reaches, based on the dominant erosion types:

Reach I: upstream of point A (i.e., the confluence site of the Quzong Tsangpo/Palong Tsangpo), with a river length of about 100 km and mean gradient of 1.5%, and the main erosion type is glacial driven.

Reach II: between points A and B (i.e. the confluence sites of the Quzong Tsangpo/ Palong Tsangpo and Guxiang Gully/Palong Tsangpo), with a river length of 80 km and mean gradient of 0.56%. It is the flattest reach of the Palong Tsangpo River and once had obvious erosion.

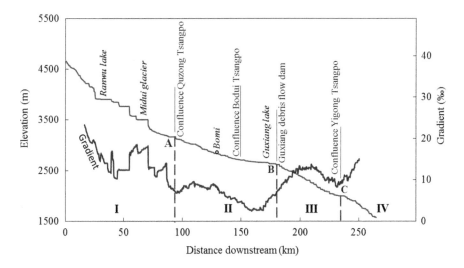

Figure 2.19 Longitudinal profile and channel gradient along the Palong Tsangpo River.

Reach III: between points B and C (i.e. the confluence sites of the Guxiang Gully/Polong Tsangpo and Yitong Tsangpo/ Palong Tsangpo). This reach is prone to debris flows.

Reach IV: downstream of Point C (i.e. the confluence site of the Yigong Tsangpo/Palong Tsangpo), with a mean gradient of 1.3%. Avalanche is the main erosion type in this reach.

This section focuses on Reach II, where glacial erosion is the dominant erosion type.

Numerous small scale glaciers have developed along Reach I of the Palong Tsangpo River. The glaciers slowly move downwards to the river and transport a massive amount of glacial morainic materials. The glacial moraine may be stabilized in the river channel if streambed structures (e.g., clusters or step-pools) form in the channel (Fig. 2.20). A large amount of glacial till from sudden glacial movements may choke the river channel and form stable morainic dams. The Ranwu Lake on Reach I is a lake formed by a morainic dam (Fig. 2.19). The distance from the Ranwu Lake to the Midui Glaciers is about 40 km. Six large glaciers distribute along this reach, and morainic dams have developed on all the confluence sites of the glaciers and the Palong Tsangpo River channel. The flow velocity upstream of the morainic dams is slow. Step structures have formed downstream of these dams in reaches with a longitudinal length of 100–800 m and mean gradient of 6%. Some of these step structures have up to 7 steps with step heights of 0.8–2.2 m (Fig. 2.20).

The fluvial morphology of Reach II has been seriously influenced by the Guxiang Gully, which is located at the downstream end of Reach II (Fig. 2.19). A massive amount of materials from the Guxiang Gully are transported into the Palong Tsangpo River during frequent debris flow events. About 20 debris flow events occur in the Guxiang Gully each year, with an average erosion amount of 1–6 million m³ transported each time. The massive amount of materials transported into the river channel

Figure 2.20 Streambed structures formed on glacial moraine (a) massive amount of morainic materials from glaciers; (b) the Ranwu Lake was formed due to a morainic dam; (c) step-pool structures develop on morainic deposits; and (d) cluster structures form on morainic deposits.

gradually form a great knickpoint, causing deposition upstream and incision downstream. Sediment has deposited in the lake and formed wide valleys (the maximum valley width amounts to 1,500 m and the maximum channel width is 800 m). The planform channel pattern is braided, with the maximum braiding index amounts to 7.

Reach III is located at downstream of the Guxiang Lake. Several debris flow gullies (including the Tianmo Gully) distribute along Reach III. The banks of Reach IV is quite steep and avalanches are prone to occur. Field investigations on a channel with a total length of 10 km found eight large-scale avalanche sites.

2.2.1.2 Two types of glaciers

Results from remote sensing data have shown that, the glacier area in the Palong Tsangpo basin amounts to approximately 36% of the total basin area (Fig. 2.18). Field investigations on these glaciers have shown that the main types of the glaciers are **hanging glacier** and **valley glacier**. Hanging glaciers distribute along the river. The glacial valley slope is about 30 degree and its cross sections often have deep "V" or "Υ" shapes. No obvious ice tongues or integral morainic structures have developed. Valley glaciers have obvious ice tongues and snow grain basins (accumulation areas). A massive amount of moraine materials have been transported downstream because of the slope rocks cut by glacier. These moraine materials have piled up downstream

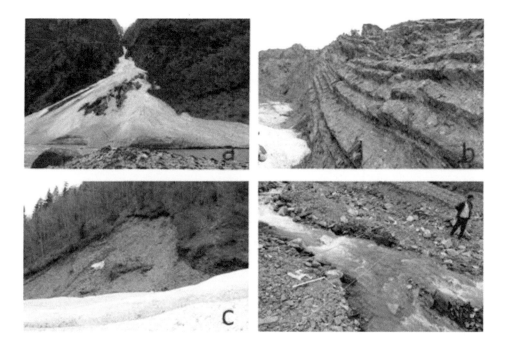

Figure 2.21 Hanging glacier and glacial erosion (a) typical hanging glacier valley; (b) glacial movement cutting the bedrock; (c) glacier erosion on bank slope; and (d) melted flow discharge measurement at downstream part of the chosen glacier.

of the glaciers and formed complete lateral moraines and final moraines. The moraine dams which have formed downstream from the valley glaciers in the Palong Tsangpo basin are 50–200 m high. The moraine materials not only change the local relief and morphology, but also have significant impacts on fluvial processes.

Many **hanging glaciers** have developed along the upper reaches of the Palong Tsangpo River. A typical hanging glacier is shown in Fig. 2.21 (a). Under the effects of glacial erosion, the glacial valley take an obvious "⑂" type cross section. The lower slope is quite steep with an average gradient of 75°, indicating an accelerated incision rate. The longitudinal profile of the glacier show a downward concave form. The upstream most section has a gradient of 31° with an elevation drop of 310 m. The middle section has a gradient of 29° with an elevation drop of 110 m. The downstream section has a gradient of 25° with an elevation drop of 203 m. The volume of the accumulated deposit was estimated to be approximately 140×10^3 m^3. The size distribution of the sediment accumulated on the glacier is shown in Fig. 2.22.

The Midui Glaciers were chosen as typical cases for field investigations conducted by the authors. Glaciers densely distribute along the Midui River (a left bank tributary of the Palong Tsangpo River). Eight glaciers have developed in a reach of about 2 km long along the Midui River. Glacial movement has a strong cutting force on bedrocks (Fig. 2.21 (b)) and erosion effects on bank slopes (Fig. 2.21 (c)). One glacier was chosen for detailed measurements (Fig. 2.21 (d)). The glacier was divided into 2 sections, the upper section (5000–3900 m) and the lower section (lower than 3900 m). The upper

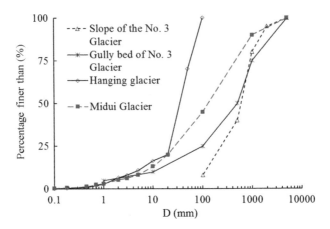

Figure 2.22 Size distributions of the sediment eroded and transported by glacier.

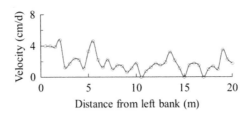

Figure 2.23 Mean velocity distribution of glacial movement measured by calculating the movement distance of the gravels.

section was about 490 m long and quite narrow (width < 10 m), while the lower section was wider (210 m) and its fan-shape spreading angle amounted to 26°, with a mean longitudinal gradient of 31%. The mean deposit depth was estimated to be 35 m, based on the gradient difference between the glacier slope and the hill slope, and the volume of the glacier deposit was about 1.8 million m³.

Two field investigations were conducted in 2012 to measure the velocity of glacial movement. The first investigation was conducted on May 12, and the second was conducted on May 16. A cross sectional line of 51 m long was put on the glacier at elevation of 3985 m. A length of 20 m in the middle part of the line was chosen for measurements, and stones (gravels) every 0.5 m interval were put on the line to mark possible location changes due to glacial movement. The movement distances of the stones (gravels) were measured and the mean moving velocity was calculated. The measured glacial movement velocity is shown in Fig. 2.23, with the mean velocity being 1.9 cm/d. The discharge formed due to glacial movement was about 0.0007 m³/s, based on the glacier width and depth at the measured cross section.

There is a small creek at the very downstream of the glacier chosen for the measurements, where flow is recharged by melting water from the glacier. Two measurements were conducted (Fig. 2.21 (d)) at 1:00 pm on May 12 and at 11:00 am on May 16,

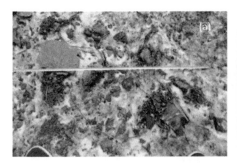

Figure 2.24 (a) Glacial movement transports a massive amount of materials to the river channel; and (b) Layer sequences form during glacier accumulation.

respectively. The discharges were determined to be 0.191 and 0.037 m³/s, respectively. The two measured discharges differed significantly, and were generally 100 times of the discharge estimated based on glacial movement, indicating that the glacier was melting.

The main motion form of a hanging glacier is typically bursting and siltation. When there is a large snow fall, snow and other materials which have accumulated in the snow field can suddenly lose their stability and move into the glacier valley with a motion distance of over 1 km. Even though sediment particles may distribute in different layers of the glacier, the episodic motion often causes a very high sediment concentration in the surface layer, and, hence, transports a huge amount of materials (Fig. 2.24(a)). Eleven 0.5 m × 0.5 m zones were chosen to measure the total sediment load on the glacier surface. The results showed that the surface sediment concentration ranged between 4.3–8.2 kg/m². The vertical profile (Fig. 2.24(b)) of the glacier was layered with the depth of each layer being about 0.2 m. The average layer depth was 12–30 cm, and the estimated sediment concentration was 22–41 kg/m³. Then the annual transport rate of sediment from this glacier was estimated to be roughly 700 ton, based on the measured glacial movement velocity, cross sectional scale and sediment concentration. If the glacier density is considered, the sediment transport load per 1 km² from the glacier may reach up to 5,400 ton since there are about 8 glaciers/km² which have developed in this region.

Many **valley glaciers** also distribute in the Palong Tsangpo basin. Valley glaciers often develop on higher elevations and have larger sizes in scale, compared to hanging glaciers, and the volume of the morainic deposits is very large. Part of these deposits will be transported into the fluvial system as the glaciers melt. Sometimes, valley glaciers move downstream directly into the rivers and block the river channels. Fig. 2.25 shows four valley glaciers (named as No. 1, 2, 3 and 4) in Bomi County and near the Galongla Tunnel, a newly-built road tunnel linking Bomi and Motuo counties (the latter was once the only county in China which was isolated from the national highway system). The length and area of the four glacial basins are shown in Table 2.3.

The snow grain basin of the No. 1 glacier (Fig. 2.26) is located at elevation 4500 m. The morainic deposits have formed drumlins with obvious layers. The geometric form of the sediment particles is basically angular, composed mainly of slate and phyllite, and

Figure 2.25 Four valley glaciers in Bomi County near Galongla tunnel in the Palong Tsangpo basin.

Table 2.3 Basic features of the 4 glaciers near Galongla Tunnel.

No.	Length (km)	Area (km²)
1	2.5	11.5
2	3.9	10.0
3	1.4	2.8
4	1.0	1.5

less marble. Stones in this area are covered by moss. Young trees with circumferences of about 25 cm have grown on the deposits. Lateral moraine dikes have naturally developed along both sides of the glacier, with heights of approximately 55 m and slopes of 33 degree. Downstream of the glacier is a flat and wide outlet (500 m in width) which formed during flood flashing. The longitudinal profile of the glacier is shown in Fig. 2.27. The No. 1 Glacier is slowly retreating in general, and the end moraine can be seen from the end of the glacier (Fig. 2.26(a)). There is a lama temple built near this glacier, and lamas often put stone piles at the end of the glacier for the sake of praying. The retreating velocity of the glacier in the past 150 years was estimated to be about 7 m/year, based on the records of the location change of these stone piles in different years.

The No. 2 glacier (Fig. 2.28) is also retreating, but more seriously and has been in movement for many years, according to the lamas from the nearby temple. A large amount of morainic deposits can be seen on the surface of the glacier.

Detailed field surveys were conducted on the accumulated deposits of the No. 3 Glacier. Compared to the No. 1 and 2 glaciers, the No. 3 Glacier is much more active. A very large ice fall can be seen from Fig. 2.29. The slope of the glacier deposit is roughly 30°. The glacier deposit is approximately 210 m high, and the azimuth of the glacier is 100 degree. Pine trees grow on the bottom and middle parts of the deposit.

Figure 2.26 No. 1 Glacier near the Galongla Tunnel (a) end of the glacier; and (b) end moraine with different layers.

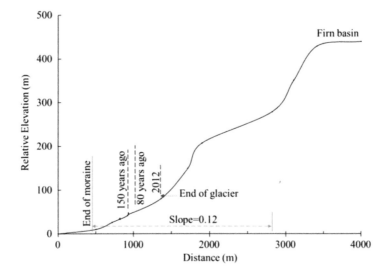

Figure 2.27 Longitudinal profile of the No. 1 Glacier.

The trunk circumferences of the trees range between 34–72 cm. Tree core samples showed that the ages were 27–29 years for the trees growing on the bottom of the deposits and 14–17 years for those on the middle part of the deposits. The stones on the bottom of the deposits are very large (see size distribution in Fig. 2.22), with moss growing on them. Stones on the middle part of the deposits are smaller and only exposed to air for a short time, hence almost no moss or other vegetation have developed on them. From Fig. 2.30(a), it can also be seen that big trees are weighed down by large stones, indicating that the morainic deposits came to this area recently and the glacial movement is quite active.

The cross sectional profile of the moraine deposit shows a "U" shape (Fig. 2.30 b), which is actually the "gully bed" of the glacier. In winter season, moraine materials

Figure 2.28 No. 2 Glacier near the Galongla Tunnel.

Figure 2.29 A very large ice fall on the No. 3 Glacier.

move down from the cornice glacier, bringing a large amount of morainic deposits. The moraine materials move along the previous deposits and form a new ice channel ('ice river'). The size of the deposits on the gully bed (ice channel) is smaller than those on the slope (Fig. 2.22).

It's quite dangerous to walk on such morainic deposits since there are many ice remnants which are not completely melted yet and prone to form ice cracks and holes. The slope of the No. 3 hanging glacier is 40–45°. Big stones sometimes fall down along the valley glacier and produce strong bloomy sound. To estimate the volume of the glacial morainic deposits, one longitudinal profile and four cross sectional profiles were arranged on the morainic deposits of the ice channel. The elevations of 312 points and 472 points were surveyed at two separate times, one in September 2012 and the other in April 2014, respectively. The digital elevation terrain figures and cross sectional profiles generated based on the two surveys are shown in Figs. 2.31 and 2.32.

Figure 2.30 Different parts of the morainic deposits of the No. 3 Glacier (a) Middle part (the man in red clothe is 173 cm tall, direction to downstream); and (b) Upper part (direction to upstream).

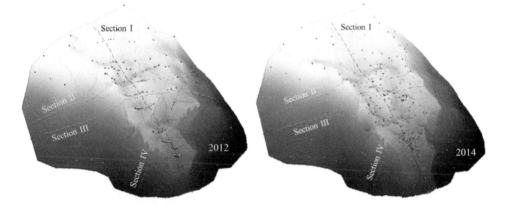

Figure 2.31 Digital elevation terrains of the morainic deposits of the No. 3 glacier based on field surveys in 2012 and 2014.

The volume of the morainic deposits of the No. 3 Glacier, based on the field survey, is about 0.11×10^9 m^3. The surveys in September 2012 and May 2014 showed that the volume increased by about 0.6–1.0 million m^3 after two active years. Then the average glacial erosion rate was estimated to be roughly 0.18–0.3 million ton/km^2/year, based on the glacial area (2.8 km^2) and void ratio (assumed to be 35%).

2.2.2 Effects of glaciers on fluvial morphology

Two important glaciations (glacial periods) occurred on the plateau during the Quaternary period, one is called the Guxiang Glaciation (about $113 \sim 136$ ka BP), and the other one is called the Dali Glaciation (about $10 \sim 70$ ka BP) (cf. TECAS, 1986; Li, 1992; Zhou et al., 2007). The intensity of glacial movement has significantly reduced since the two glaciations. Modern glacier only partially affects fluvial morphology.

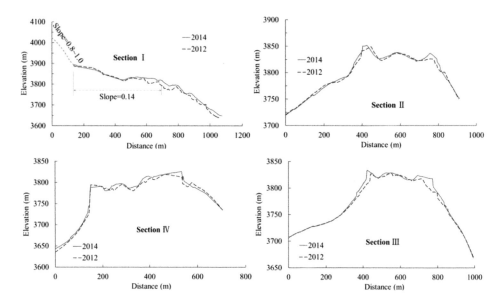

Figure 2.32   Longitudinal (Section I) and Cross sectional (Sections II, III, and IV) profiles of the morainic deposits of the No. 3 Glacier.

Figure 2.33   A historical moraine dam near the confluence of the Quzong Tsangpo/Palong Tsangpo is still well-preserved.

During the Quaternary glacial period, the spatial range and intensity of glacial movement was much larger (and higher) than the present time, as glaciers at that time even moved into the Palong Tsangpo River channel and formed massive ice moraine dams. Some of those ice moraine dams are still well-preserved even till today, as shown in Fig. 2.33. Field surveys were conducted in the Bodui and Quzong Tsangpo rivers, two major tributaries of the middle and upper Palong Tsangpo River, to examine the

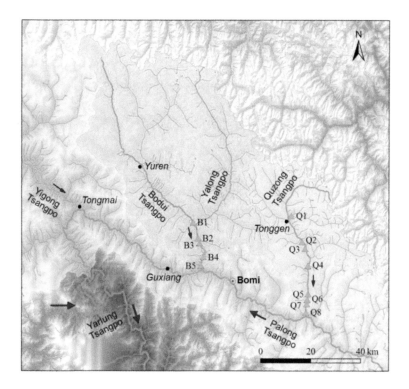

Figure 2.34 Field investigation sites on the Bodui and Quzong Tsangpo rivers.

effects of morainic deposits on the evolution of fluvial morphology. The survey points are shown in Fig. 2.34.

2.2.2.1   Quzong Tsangpo River

Under the influence of ancient glaciers, the valley bottom of the Quzong Tangpo River is as wide as 1 km in some reaches, and the braiding index is over 5, even though the Quzong Tsangpo is just a small mountain river. The channel along the wide valley reach is very flat, with sand being the main sediment component (Fig. 2.35). Fig. 2.36 shows the longitudinal, gradient, channel width and braiding index variations along the river.

The size distributions of the survey points on the Quzong Tsangpo River are shown in Fig. 2.37. The authors of this book used eye estimation for sediment particles larger than 100 mm, sieved sediment with size between 1–100 mm, and used laser scattering particle analyzer (MasterSizer 2000) to analyze the sediment particles with size <1 mm. The ancient glaciers had significant influence on the composition of the channel bed materials. Different from the general case that the sediment becomes finer along the river reach from upstream to downstream, rather, the sediment size is larger near the tributary gully mouth where the river is affected by the development of the ancient glaciers (Q5, Q7), and the sediment size is smaller upstream of the tributary gully mouth (Q4, Q6).

Figure 2.35 Channel near survey point Q3. The valley is quite wide (on the order of 500 m) and the channel bed is flat, with sand being the main component of the river bed.

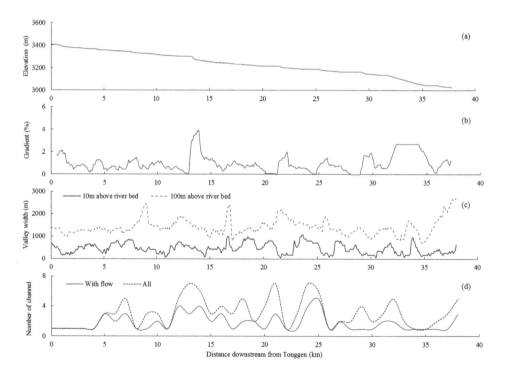

Figure 2.36 Variations of longitudinal profile, gradient, channel width and braiding index along the reach of the Quzong Tsangpo.

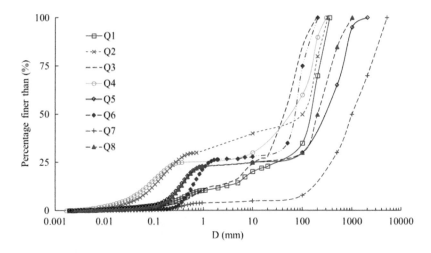

Figure 2.37 Size distributions of sediment along the channel bed of the Quzong Tsangpo.

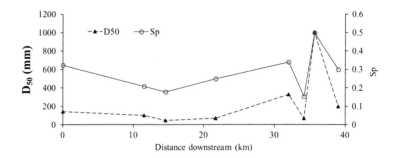

Figure 2.38 Variation of the median size (D_{50}) and streambed development degree (S_P) along the Quzong Tsangpo River.

The variation of development degree of streambed structures, S_P (cf. Wang et al., 2004; 2009 for calculating method) and the median size (D50) of the channel bed sediment along the river are shown in Fig. 2.38. The two indices vary similarly, with the value of S_P varying more gently. In general, the sediment composition of the channel bed is finer and the streambed structure is weaker for the reaches upstream of the glacial dam; while the sediment composition is coarser and the streambed structure is stronger for the reaches downstream of the glacial dam. The reaches corresponding to the peak values of S_P and D50 are those affected by glaciers, among which the glacier at survey point 4 had the largest scale.

2.2.2.2   Bodui Tsangpo River

The glacier development during the Guxiang and Dali Glaciations in the Quaternary period also markedly influenced the fluvial morphology of the Bodui Tsangpo River. A complete piedmont glacier once developed along the Bodui Tsangpo River during the

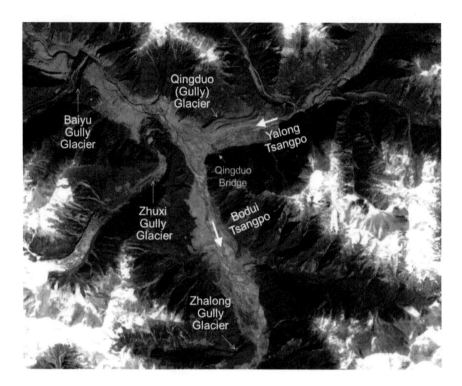

Figure 2.39   Quaternary glacier remnants in the lower reaches of the Bodui Tsangpo River (source: Google Earth).

Guxiang Glaciation, and the glacier along the main stem Bodui Tsangpo River entered into the Palong Tsangpo River and once transported downstream to near the Guxiang Lake. The glacier scale was relatively small during the Dali Glaciation, however, it still moved along the main stem of Bodui Tsangpo and once reached the mouth of the Baiyu Gully (Fig. 2.39). The remnants of the ancient morainic deposits in the Baiyu Glaciation[1], one of the sub-glaciation periods during the Dali Glaciation, were well preserved due to the short time (about 10,000 years). It can be seen clearly from Fig. 2.39 that the remnants of four large ancient glaciers, i.e., the Baiyu Gully Glacier, Zhuxi Gully Glacier, Qingduo Glacier, and Zhalong Gully Glacier, all extended to the Bodui Tsangpo River and formed lateral moraine dams with elevations about 200 m higher than the current river bed.

A terrace on the north side of the Qingduo Bridge (30°03′21.0″N 95°35′54.1″E), which is located on the Yalong Tsangpo River (and just upstream of the confluence site of Yalong Tsangpo/Bodui Tsangpo), was formed by morainic deposits from the

---

[1] Baiyu Glaciation $((11.1 \pm 1.9) \sim (18.5 \pm 2.2)$ ka BP) is considered to be the third of four sub-glaciation periods during the Dali Glaciation on the Plateau. cf. TECAS, 1986; Li, 1992; Zhou et al., 2007.

glacier moving along the Qingduo Gully (Fig. 2.39). OSL (optically stimulated luminescence) samples were taken from the terrace, and the dating results showed that the development time of the terrace could trace back to 39.4 ± 10.3 ka BP, indicating that the development period was the second period of the Dali Glaciation, which corresponded to MIS (Marine isotope stages) 3b of the deep sea isotope period (roughly ranged between 40–56 ka BP) (Zeng et al., 2007; Owen and Dortch, 2014) and was earlier than the Baiyu Glaciation.

Although the spatial scale of glaciers developed during the Guxiang Glaciation was quite large, the morainic deposit remnants were poorly maintained due to its old age and severe gravitational and water erosion. Consequently, only a small numbers of moraine boulders on the high terraces along some of the tributaries were well preserved. In contrast, the morainic deposits from the Baiyu Glaciation were better maintained.

Based on previous analyses, piedmont glaciers mainly developed in the Bodui Tsangpo basin during the Guxiang Glaciation. Following the Guxiang Glaciation was an interglacial period which lasted for tens of thousands of years. The morainic deposits in river valley were likely eroded during that period, but the entire upstream channel of the Bodui Tsangpo River was covered again with glacial morainic deposits during the subsequent Dali Glaciation (including Baiyu Glaciation). Furthermore, the main channel was also dammed by the Baiyu Gully Glacier and Zhuxi Gully Glacier from the tributaries and lakes formed in the upstream channel, and then moraine materials deposited in the lakes and made the valley bottom wider and wider. The depth of deposition was estimated to be over 150 m based on field survey of the lateral morainic dams on the two tributaries. During the subsequent ten thousand years, which lasted from the Baiyu Glaciation to present time, the upper reaches of the Bodui Tsangpo River have been continuously incised from the loose fluvial and lacustrine deposits. The current incision depth in the upper reaches is over 70 m, and the valley width variations are presented in Fig. 2.40 (c).

The Bodui Tsangpo River in a whole has a quite wide valley because it was filled up by morainic deposits during the Guxiang Glaciation. Since the glaciers in the Baiyu Gully and Zhuxi Gully were near the river main stem, the morainic deposits were prone to enter the Bodui Tsangpo River and caused deposition in the upper reaches, hence the channel width, incision depth and planform patterns of current channel bed are quite different in the upstream and downstream reaches (the separation point is near the confluence site of Zhuxi Gully/Bodui Tsangpo, at the location of 33 km on the horizontal axis of Fig. 2.40).

Two reference elevations are used to define the valley width of the Bodui Tsangpo River: one is the elevation 100 m higher than the water surface, and the other is 10 m higher than the water surface. The valley width at the first reference elevation (100 m) is more or less consistent from upstream to downstream along the river; while the valley width at the latter reference elevation (10 m) shows a significant difference from upstream to downstream with small values in the upstream reaches and large values in the downstream reaches, and the variation is more significant in the downstream reaches. The difference in the valley width at the two reference elevations is smaller in the downstream reaches compared to the upstream reaches, and in the same variation trend. The valley width at the first reference elevation represents the width of the glacial valley which developed due to glacial movement during the Guxiang Glaciation; while

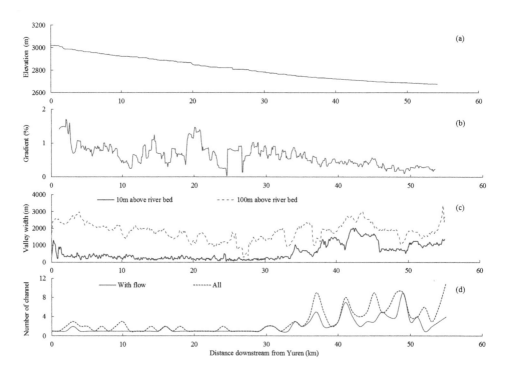

Figure 2.40 Longitudinal, gradient, channel width and braiding index variations along the reach of the Bodui Tsangpo.

the second valley width represents the current channel width, and means that the upper reaches has been in incision from previous glacial moraine deposits.

The cross sectional profiles in the upstream and downstream reaches can also be used to evaluate the change of the river valley along the river channel. For example, Fig. 2.41 shows two typical cross sectional profiles in the upstream and downstream reaches of the Bodui Tsangpo which are 35 km apart and have similar cross sectional geometries in general and are both classified as typical U-form valleys. The main difference between the two profiles is that the upstream section has a higher elevation and its river channel has been incised by more than 100 m on its U-form valley. Other cross sectional profiles in the upstream reaches are similar to that in Fig. 2.41, and the incised depth range between 70–140 m. No obvious incision is noticed in the sections in the lower reaches, and the elevation of the wide valley in the upper reaches is much higher than that in the lower reaches. The braiding index (Fig. 2.40 d) also shows that, the planform pattern in the upper reaches is single-thread since the channel has been experiencing incision, while the planform pattern in the lower reaches is primarily braided with multiple channels.

The lower reaches of the Bodui Tsangpo River can also be divided into two sections, i.e., the upper and the lower sections. The upper section is the reach between the confluences of Zhuxi Gully/Bodui Tsangpo and Zhalong Gully/Bodui Tsangpo, while the lower section is the reach from the confluence of Zhalong Gully/Bodui Tsangpo to

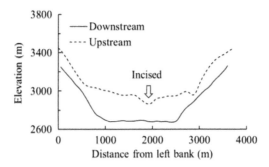

Figure 2.41 Typical cross sectional profiles in the upper and lower reaches of the Bodui Tsangpo.

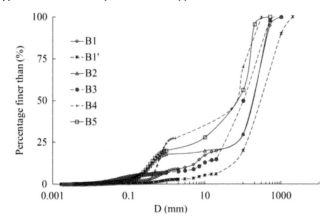

Figure 2.42 Size distributions of sediment at different surveying locations of the Bodui Tsangpo.

the mouth of the Bodui Tsangpo River. The planform pattern of the upper section is braided. There is a terrace with a height of about 1 m above the current river bed, and the reason may be that the river was dammed by morainic deposits from the Zhalong Gully. The planform pattern of the lower section is much more braided with many unstable channels. The river bed is still in deposition, and its planform pattern is the same as that of the Palong Tsangpo River which the Bodui Tsangpo River empties into. Hence, the deposition in the lower section of the Bodui Tsangpo and in the lower reaches of Palong Tsangpo may be caused by a single event, and the occurrence of this event was likely relatively recent based on the deposition condition.

The size distributions of the survey points in Bodui Tsangpo are shown in Fig. 2.42. The same method was used for the size composition analysis as that for the Quzong Tsangpo (Fig. 2.37). Although the sampling sites were all located in the lower reaches of the river, the river bed sediment composition still show significant differences between the upper and lower sections. The river bed is mainly composed of cobbles in the upper section with bigger sizes for the largest particles, while for the lower section the bed is composed of 'dual structure' which is located near the deposition reach of the old Guxiang Lake, and the sediment sizes of the largest particles are relatively small.

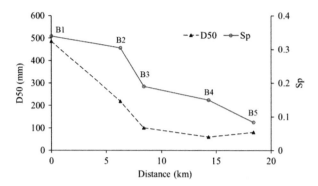

Figure 2.43 The variation of median size (D50) and streambed development degree (S_P) along the Bodui Tsangpo.

The variation of S_P (cf. Wang et al., 2004; 2009) and the median size (D_{50}) of the channel bed sediment along the river is shown in Fig. 2.43. The two indices vary similarly. In general, the sediment size of the channel bed becomes finer and the streambed structure becomes weaker along the river reach. Fig. 2.43 also shows the difference of S_P and D50 for the reaches between confluences of Zhuxi Gully/Bodui Tsangpo to Zhalong Gully/Bodui Tsangpo (B1, B2 and B3) and between the confluence of Zhalong Gully/Bodui Tsangpo to the Bodui Tsangpo river mouth (B4, B5).

2.2.3 Impacts of glaciers on the Palong Tsangpo

The Bodui and Quzong Tsangpo rivers are not far away from each other and the distance between the two confluence sites of the two rivers with the main stem Palong Tsangpo River is just 50 km, hence the two river basins share similar climate and environment. The glacial valleys of both the Bodui and Quzong Tsangpos are in typical U-form. It is common in these rivers that large boulders were transported due to glacial movements. The moraine deposits in some branch gullies are very well kept.

The development period and intensity of glaciers in this region differed markedly from other high-altitude or west-wind region mountainous glaciers in China due to the strong uplift processes and unique climate. The glaciers on the plateau during the Guxiang Glaciation and Dali Glaciation had much larger spatial scale than those existing today, however, many researchers pointed out that the ancient glaciers on southeast plateau were isolated piedmont glaciers which distributed along the river reaches rather than forming an intact continental ice cover (TECAS, 1983; 1984; 1986; Li, 1992; Zeng et al., 2007; Owen and Dortch, 2014).

The main stem of the Palong Tsangpo River has been dammed by glaciers from its tributaries at many sites. Glaciers came into the Palong Tsangpo River and moved downstream along the river channel until its moraine end reached the Guxiang Lake. The spatial scale of the glaciers during the Dali Glaciation was relatively small, however, many glaciers still moved into the Palong Tsangpo River since they were very close to the main stem river.

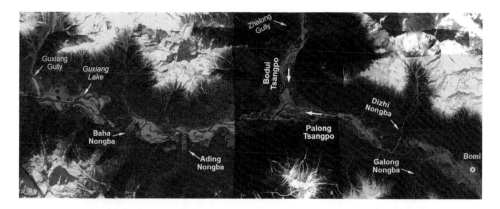

Figure 2.44 Wide valley reaches upstream of the Guxiang Lake and near Bomi County (source: Google Earth).

Since the Palong Tsangpo basin is located in the southeast edge of the plateau, a region with the highest uplifting rate since Quaternary, the cross sectional profiles of most river reaches in this region are 'V'-shaped or have steep cliffs. However, the reach from Bomi to the Guxiang Gully is unusually wide (Figs. 2.12 and 2.44) with the width ranging between 500–1,500 m. This wide reach can be divided into at least three sub-reaches based on the channel geometry, incision depth, channel bed elevation, deposition status and cross sectional profiles. From upstream to downstream, the three sub-reaches are: upstream of Bomi to Galong Nongba[2] Gully (it is named the **Bomi ancient lake** in this book), Dizhi Nongba Gully to Baha Nongba Gully (**Bodui ancient lake**), and the **Guxiang Lake** (Figs. 2.44 and 2.46).

The **Bomi ancient lake** is 8 km long. The deposition in this sub-reach started a long geological time ago. Bomi County is located in this sub-reach. The channel of the Palong Tsangpo River near the Bomi ancient lake is now in incision and has a single-thread planform. It has been incised by about 20–40 m from lake deposits. Downstream of the Bomi ancient lake is the Galong Nongba Gully, a left-bank tributary gully which was shaped by ancient glacier (Fig. 2.44). Now morainic deposits with a height of about 80 m is still preserved in the Galong Nongba Gully. Downstream of the Galong Nongba Gully is the Dizhi Nongba Gully, a right-bank deris flow gully with a large deposit fan intruding into the Palong Tsnagpo River. Some studies have shown that the glacier in the Galong Nongba Gully might have developed during the last glaciation, and the Bomi ancient lake most likely formed after the Galong Nongba Glacier had dammed the channel of the Palong Tsangpo River (e.g., TECAS, 1984; 1986).

Similar fluvial processes also include the formation of the Songzong ancient lake which is located upstream of the confluence site of Quzong Tsangpo/Palong Tsangpo, the formation of the wide valley of the Quzong Tsangpo River, and the deposition of the Bodui Tsangpo River channel upstream of the confluence site of Zhuxi Gully/Bodui Tsangpo. The deposition of these wide valleys all occurred in the last ice age (Baiyu

---

[2]Nongba means creek or gully in Tibetan.

Figure 2.45 Lateral moraine dam formed by ancient glacier near the confluence site of Ading Nongba/Palong Tsangpo (camera to the direction of the east).

Glaciation) when the main river channels were dammed by glacier morainic deposits from tributary gullies. Now the river channel in the Songzong ancient lake has been incised by up to 70 m, and the channel on the historical deposition body of the Bodui Tsangpo River has also been incised by about 70–140 m. Although the Dizhi Nongba Gully is similar to the Guxiang Gully as a large debris flow gully, the active period of the Dizhi Nongba Gully is much earlier than that of the Guxiang Gully. Hence, the formation of the Bomi ancient lake was also possibly caused by debris flow damming, which was similar to the formation of the Guxiang Lake.

The **Bodui ancient lake** starts from the confluence site of Baha Nongba/Palong Tsangpo and extends upstream along the Palong Tsangpo River to the Dizhi Nongba Gully, and along the Bodui Tsangpo River to the Zhalong Gully (Fig. 2.44). The Bodui ancient lake has a length of 26 km along the Palong Tsangpo River and 6.5 km along the Bodui Tsangpo River. Both end points of the lake deposition are at elevations about 2670 m. The channel along the Palong Tsangpo River is highly braided and wandering. The current river channels spread out at the entire valley bottom, and are still in deposition. Although sand bars in the river channel are typically not stable, vegetation (trees and brush) still develops on some bars. In the lower part of the Bodui ancient lake, two gullies, Ading Nongba and Baha Nongba, have developed. It can be deduced from the local relief map that: the width of the glacial valleys of the two gullies is over 300 m; the piedmont lateral morainic dams are about 200 m higher than the ancient lake; and near the confluence site, the lateral dams are about 90 m higher than the water surface of the river, and the dam heights satisfied the requirements to form the Bodui ancient lake. The scale of the two glacial valleys is almost the same as that of the Dongqu glacial valley, which was formed by a glacier which dammed the Palong Tsangpo River and formed the Songzong ancient lake during the Baiyu Glaciation. The summit elevations of the Ading and Baha Nongba gullies are both around 4500 m. Very large ice cirques have developed in both gullies, and modern

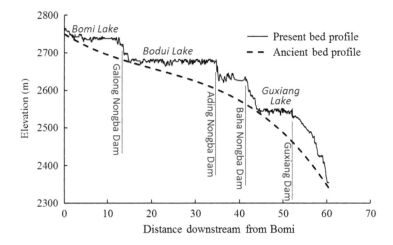

Figure 2.46 Longitudinal profile of the Yarlung Tsangpo River reach downstream from Bomi and the estimated old river profile before damming incidents.

glaciers are still developing in the upstream reaches of the gullies. Therefore, the Bodui ancient lake most likely formed when glacial morainic deposits moved into the Palong Tsangpo River and dammed the channel during a certain period. Fig. 2.45 shows the lateral moraine deposit dam which was formed by an ancient glacier at the confluence site of Ading Nongba/Palong Tsangpo.

The Palong Tsangpo River channel downstream of Baha Nongba is quite narrow with a length of about 2 km and an elevation difference of about 50 m (Fig. 2.44). Downstream of this reach is the third wide reach, the Guxiang Lake, which formed due to the damming by the glacial debris flow from the Guxiang Gully which occurred in September 1953. A huge amount of materials from the debris flow dammed the Palong Tsangpo River and caused the water level upstream of the dam to rise by about 40 m. Several hundred debris flows occurred after that, and the current Guxiang Lake with a length of 6 km and a width of 1–2 km has formed. The deepest part of the lake has a water depth of about 20 m, according to field investigations conducted by Chinese Academy of Sciences in 1973–1980 (TECAS, 1983; 1984).

The influence of ancient glacial processes on the longitudinal profile of the downstream rivers can be clearly seen from Fig. 2.46, the longitudinal profile of the Palong Tsangpo River reach downstream from Bomi. The solid line on the figure was extracted based on ASTER DEM (spatial resolution 30 m), and the dashed line presents the longitudinal profile before the damming incidents which was estimated based on the overall longitudinal profile of the Palong Tsangpo River and the elevation difference of the wide valley reach. Based on this figure, the longitudinal profile of the Bodui ancient lake has two stairs, which was likely caused by separate damming incidents of the Ading Nongba and Baha Nongba Gullies. The elevation differences of the downstream end points of these ancient lakes and the old longitudinal profile before damming incidents correspond to the heights of those dams. The heights of the lateral moraine dams formed by ancient glaciers correspond to Fig. 2.46 quite well. Although the summit period of the last Glaciation in the Quaternary period has been more than tens of thousands of years ago, the effects of the ancient glaciers (Dizhi, Ading and Baha Nongbas) on the fluvial morphology evolution have been everlasting, and, to some extent, the effects are decisive.

2.3 STREAMBED STRUCTURES CONTROLLING RIVERBED INCISION

2.3.1 Streambed structures

Streambed structures are boulders and cobbles on mountain streambeds which are rearranged by flood flows to reach high resistance and high bed stability. In general, these structures are often in the form of sediment clusters or aggregations. Streambed structures inhibit bed particle movements (Laronne and Carson, 1976; Strom et al., 2004; Wittenberg and Newson, 2005; Oldmeadow and Church, 2006; Wittenberg et al., 2007) and thus significantly influence bed load transport dynamics (Church et al., 1998; Hanssan and Church, 2000; Lamarre and Roy, 2008). Hence, streambed structures play an important role in mountain streams through controlling bed incision and maintaining channel stability.

Step-pool system is the most important type of bed structures (Wang and Zhang, 2012; Wang et al., 2014). The steps are generally composed of cobbles and boulders, which alternate with finer sediments in pools to produce a repetitive, staircase–like longitudinal profile in the stream channel. The tight interlocking of particles in each step gives the structure an inherent stability that only extreme floods are able to disturb. Step-pool systems typically develop in high gradient streams. Other bed structures, such as ribbing structures and cobble or pebble clusters, develop in low gradient mountain streams (Wang et al., 2014; de Jong, 1991; Wittenberg, 2002). Many studies have found that step-pool systems maximize flow resistance (Abrahams et al., 1995; Rosport, 1997; Strom and Papanicolaou, 2007, Lee and Ferguson, 2002; Wilcox et al., 2006) and dissipate flow energy by promoting tumbling flow and roller eddies (Wohl and Thompson, 2000). Thus, step-pool systems dissipate a huge amount of energy, and protect the channel bed from erosion. The authors have performed experiments and field measurements and found that the energy dissipation ratio by step-pools is more than 60% (Wang et al., 2004). Bed load motion consumes flow energy as well. Gravel particles slide or move by saltation on the stream bed, applying a pressure on the stationary bed materials. With enough bed load, the channel bed is not scoured and the channel incision is, therefore, stopped. Bed structures and bed load motion have similar functions in that they both consume flow energy and control stream bed incision (cf., Wang and Zhang, 2012).

2.3.2 Equivalency principle of bed structure and bedload motion

The principle of equivalency of bed structures and bed load motion is particularly important for mountain streams. In laboratory experiments and alluvial rivers with bed sediment composed mainly of gravel, sand and silt, there are no bed structures. Bed load transportation depends on the shear stress of the flow. In mountain streams, however, bed structures can be very strong and consume most of the flow energy. There is no bed load motion in this case, even though the shear stress of the flow is very large.

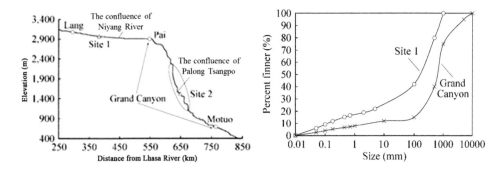

Figure 2.47 (a) Longitudinal bed profile of the Yarlung Tsangpo River; and (b) Size distributions of bed materials at Site I and Site 2 (Grand Canyon).

Figure 2.48 Bed load deposit at Site I showing clearly diagnostic features of long distance transportation: bed load particles with round shape and sorting.

Many reports have been published on the greatly reduced bed load motion due to bed structures. Yager et al. (2007) found that the rate of bed load transport in steep and rough streams depended not only on local sediment availability but also on the drag (energy dissipation) due to rarely mobile particles.

The authors of this book performed field investigations in the Yarlung Tsangpo River in Tibet and measured bed structures and depth of bed load deposits in 2009, 2010 and 2011. Figure 2.47 (a) shows one part of longitudinal bed profile of the Yarlung Tsangpo River. The abscissa is the distance from the confluence site of the Lhasa/Yarlung Tsangpo rivers. The steep section is the Yarlung Tsangpo Grand Canyon. Site 1 is located in the upstream reach of the Yarlung Tsangpo Grand Canyon, where the bed gradient is mild and the river channel is wide. Site 2 is located in the Grand Canyon near the confluence of the Palong Tsangpo/Yarlung Tsangpo rivers, where the bed gradient is high and the flow shear stress is extremely high. Figure 2.47 (b) shows the size distributions of the bed materials at Site 1 and the Grand Canyon (Site 2).

Figure 2.49   Yarlung Tsangpo Grand Canyon and bed materials at Site 2 (left); Suspended sand deposited on the bed but there was no bed load deposit at Site 2 (right).

Table 2.4   Hydraulic and morphologic features of the river at Site 1 and Site 2.

Site	S_p	Q (m^3/s)	Valley width (m)	Channel Width (m)	Gradient	P (kg/ms)	Water depth (m)	τ (N/m^2)
Site 1	0.14	9,120	1600	180	0.0009	48	10.1	94
Site 2	0.92	21,624	230	120	0.0221	3990	14.4	3128

Figure 2.48 shows the channel and bed materials at Site 1, where the bed materials had clearly diagnostic features of bed load: round shaped and sorting. The gravel and cobbles had been transported by a long distance and lost all angles and sharp edges. In the interstices of the particles, there was sand, which was transported as suspended load and deposited during flood seasons. With the instrument of EH4, which detects the depth of sediment deposits by receiving reflected electromagnetic signals from the interface of the bed rock and sediment deposit, the depth of the bed load deposit was measured to be 330 m (Wang and Zhang, 2012; Wang et al., 2015). Fig. 2.49 shows the gorge shape of the Grand Canyon and bed materials at Site 2. There was no bed load deposit because the solid particles were not in round shape and exhibited no sorting. The solid particles, large and small, had been not transported as bed load and remained their original shapes. There was suspended load deposit (sand) but no bed load deposit. Some bed particles remained stationary for a very long period of time and their surface became smooth because of abrasion from flowing water and suspended sand.

Table 2.4 lists the hydraulic and morphologic features of the river at Site 1 and Site 2, in which Q is the flood discharge with a recurrence period of 5 years. Q and the water depth were obtained from the data recorded at the hydrologic stations on the Yarlung Tsangpo River near Site 1 and at Pai (a small town at the entrance of the Yarlung Tsangpo Grand Canyon) and a station on the Palong Tsangpo River. The channel width was measured with a laser rangemeter and the bed gradient was calculated using DEM data. The unit stream power p and shear stress τ were calculated based on the flood discharge and bed gradient.

Although the rate of bed load transportation was not measured, there was obviously gravel bed load motion at Site 1 and there was no bed load motion at Site 2. The shear stress at the Grand Canyon (Site 2) was very high, which was 33 times higher than that at Site 1, but there was no bed load motion in the canyon because the bed structure dissipated most of the flow energy. The S_p value at Site 2 was 0.92, which was much higher than that at Site 1. Because the bed structure in the Grand Canyon dissipates the flow energy, the solid particles in the canyon could not be transported as bed load. For thousands of years, the gravel and cobbles bed load were transported and deposited into the upstream reaches rather than the canyon, which resulted in a large depth of bed load deposit and a wide valley width in the upstream reaches of the canyon.

REFERENCES

Abrahams A.D., Li G., and Atkinson J.F. 1995. Step-pool stream: Adjustment to maximum flow resistance. Water Resources Research, 31(10), 2593–2602.

Bagnold R.A. 1954. Experiments on a gravity free dispersion of large solid spheres in a Newtonian fluid under shear, Proceedings of the Royal Society London, 225A, 49–63.

Burbank D.W., Leland J., Fielding E., Anderson R.S., Brozovic N., Reid M.R., Duncan C. 1996. Bedrock incision, rock uplift and threshold hillslopes in the northwestern Himalayas. Nature, 379, 505–510.

Chung S.L., Lo C.H., Lee T.Y., Zhang Y.Q., Xie Y.W., Li X.H., Wang K.L. and Wang P.L. 1998. Diachronous uplift of the Tibetan plateau starting 40 Myr ago. Nature, 394, 769–774.

Church M., Hassan M.A., and Wolcott J.F. 1998. Stabilizing self-organized structures in gravel-bed stream channels: Field and experimental observations. Water Resources Research, 34(11), 3169–3179.

Coleman M., Hodges K. 1995. Evidence for Tibetan Plateau uplift before 14 Myr ago from a new minimum age for east–west extension. Nature, 3742, 49–52.

De Jong C. 1991. A re-appraisal of the significance of obstacle clasts in the cluster bedform dispersal. Earth Surface Processes and Landforms, 16(8), 737–744.

Evans S. and Delaney K. 2011. Characterization of the 2000 Yigong Zangbo River (Tibet) landslide dam and impoundment by remote sensing. In: Evans S.G., Hermanns R.L., Strom A. and Scarascia-Mugnozza G. (eds.) Natural and artificial rockslide dams. Berlin, Springer-Verlag, p. 543–559.

Glenn C.R., Kelts K. 1991. Sedimentary rhythms in lake deposits. In: Einsele G., Ricken W. and Seilacher A. (eds). Cycles and events in stratigraphy. Springer-Verlag Berlin Heidelberg, 1991, 189–221.

Harvey A.M., Wells S.G. 1987. Response of Quaternary fluvial systems to differential epeirogenic uplift: Aguas and Feos river systems, southeast Spain. Geology, 15, 689–693.

Hassan M.A. and Church M. 2000. Experiments on surface structure and partial sediment transport in gravel bed streams. Water Resources Research, 36(7), 1885–1895, doi: 10.1029/2000WR900055.

Lamarre H. and Roy A.G. 2008. The role of morphology on the displacement of particles in a step–pool river system. Geomorphology, 99(1–4), 270–279, doi: 10.1016/j.geomorph.2007.11.005.

Laronne J.B. and Carson M.A. 1976. Interrelationships between bed morphology and bed-material transport for a small, gravel-bed channel. Sedimentology, 23(1), 67–85, doi: 10.1111/j.1365-3091.1976.tb00039.x

Lavé J. and Avouac J.P. 2001. Fluvial incision and tectonic uplift across the Himalayas of central Nepal. Journal of Geophysics Research, 106(B11), 26561–26591.

Lee A.J. and Ferguson R.I. 2002. Velocity and flow resistance in step-pool streams. Geomorphology, 46, 49–71.

Li J.J. Glacial relics of monsoonal asia in the last glaciation. Quaternary Sciences, 1992, (4): 332–340. (in Chinese)

Oldmeadow D.F. and Church M. 2006. A field experiment on streambed stabilization by gravel structures. Geomorphology, 78(3–4), 335–350, doi: 10.1016/j. geomorph.2006.02.002.

Owen L.A. and Dortch J.M. 2014. Nature and timing of Quaternary glaciation in the Himalayan-Tibetan Orogeny. Quaternary Science Reviews, 88: 14–54.

Rosport M. 1997. Hydraulics of steep mountain streams. International Journal of Sediment Research, 12(3), 99–108.

Royden L.H., Burchfiel B.C., and van der Hilst R.D. 2008. The geological evolution of the Tibetan Plateau. Science, 321, 1054–1058.

Shang Y.J., Yang Z.F., Li L.H., Liu D.A., Liao Q.L., Wang Y.C. 2003. A super-large landslide in Tibet in 2000: background, occurrence, disaster, and origin. Geomorphology, 54, 225–243.

Strom K.B. and Papanicolaou A.N. 2007. ADV measurements around a cluster microform in a shallow mountain stream. Journal of Hydraulic Engineering, 133(12), 1379–1389.

Strom K., Papanicolaou A.N., Evangelopoulos N., and Odeh M. 2004. Microforms in gravel bed rivers: formation, disintegration, and effects on bedload transport. Journal of Hydraulic Engineering, ASCE, 130(6), 554–567, doi: 10.1061/(ASCE)0733-9429(2004)130:6(554).

Tapponnier P., Xu Z.Q., Roger F., Meyer B., Arnaud N., Wittlinger G., Yang J.S. 2001. Oblique Stepwise Rise and Growth of the Tibet Plateau. Science, 294(5547), 1671–1677.

Tibetan Expedition team of Chinese Academy of Sciences (TECAS). 1986. Glaciers in Tibet, Beijing: Science Press, 328p. (in Chinese)

Tibetan Expedition team of Chinese Academy of Sciences (TECAS). 1983. Tibetan geomorphology, Beijing: Science Press, 238p. (in Chinese)

Tibetan Expedition team of Chinese Academy of Sciences (TECAS). 1984. Rivers and lakes in Tibet, Beijing: Science Press, 238p. (in Chinese)

Wang Z.Y., Zhang K. 2012. Principle of equivalency of bed structures and bed load motion. Int J Sediment Res, 27(3): 288–305.

Wang Z.Y., Melching C.S., Duan X.H., and Yu G.A. 2009. Ecological and hydraulic studies of step-pool systems. ASCE Journal of Hydraulic Engineering, 134(9), 705–717.

Wang Z.Y., Xu J., and Li C.Z. 2004. Development of step-pool sequence and its effects in resistance and stream bed stability. International Journal of Sediment Research, 19(3), 161–171.

Wang Z.Y., Melching C.S., and Lee J.H.W. 2014. River Dynamics and Intergrated River Management. Springer, Verlag and Tsinghua Press, Berlin and Beijing, 847p.

Wang Z.Y., Yu G.A., Wang X.Z. Melching C.S., Liu L. 2015. Sediment storage and morphology of the Yalu Tsangpo valley due to uneven uplift of the Himalaya. SCIENCE CHINA (Earth Sciences), 58(8), 1440–1445.

Wilcox A., Nelson J.M., and Wohl E.E. 2006. Flow resistance dynamics in step-pool channels 2: Partitioning between grain, spill, and woody debris resistance. Water Resources Research, 42, W05419, doi: 10.1029/2005WR004278.

Wittenberg L. and Newson M. D. 2005. Particle clusters in gravel-bed rivers: an experimental morphological approach to bed material transport and stability concepts. Earth Surface Process and Landforms, 30(11), 1351–1368, doi: 10.1002/esp.1184.

Wittenberg L., Laronne J.B., and Newson M.D. 2007. Bed clusters in humid perennial and Mediterranean ephemeral gravel-bed streams: The effect of clast size and bed material sorting. Journal of Hydrology, 334(3–4), 312–318, doi: 10.1016/j.jhydrol.2006.09.028.

Wohl E. and Thompson D. 2000. Velocity characteristics along a small step-pool channel. Earth Surface Processes and Landforms, 25, 353–367.

Yager E.M., Kirchner J.W., and Dietrich W.E. 2007. Calculating bed load transport in steep boulder bed channels, Water Resources Research, 43, W07418, doi: 10.1029/2006WR005432.

Yin A., Harrison T.M. 2000. Geologic evolution of the Himalayan-Tibetan orogen. Annual Review of Earth and Planetary Sciences, 28: 211–280.

Zeng Q.L., Yang Z.F., Yuan G.X., Shang Y.J., Zhang L.Q., Zhao X.T. 2007. Songzong Lake: an ice-dammed lake of last glacial maximum in Palong Tsangpo River, Southeast Tibet. Quaternary Sciences, 27(1): 85–92.

Zhou S.Z., Xu L.B., Patrick M.C., David M.M., Wang X.L., Wang J., Zhong W. 2007. Cosmogenic 10Be dating of Guxiang and Baiyu Glaciations. Chinese Science Bulletin, 52(10), 1387–1393.

Zhou, C.H., Yue, Z.Q., Lee, C.F., Zhu, B.Q., Wang, Z.H. 2001. Satellite image analysis of a huge landslide at Yi Gong, Tibet, China. Quarterly Journal of Engineering Geology and Hydrogeology 34, 325–332.

Chapter 3

Meandering rivers in Sanjiangyuan

3.1 MEANDERING RIVERS

The source area of the Yangtze, Yellow, and Lancang (Mekong) rivers, situated in the northeastern Qinghai-Tibet Plateau, is named Chinese Sanjiangyuan (source of the three rivers). Meandering river pattern develops in many rivers in Sanjiangyuan, which is the source area of the Yangtze, Yellow, and Lancang rivers on the Qinghai–Tibet Plateau, as shown in Figure 1.6. Especially, the Yellow River have many meandering tributaries with extremely high sinuosity. This chapter studies the meandering rivers in Sanjiangyuan and discuss the development process and cutoff of the meanders on the plateau.

Meandering rivers are among the most prevailing river patterns in nature and have long attracted a great deal of attention from fluvial geomorphologists and hydraulic engineers because of intriguing questions over their formation, morphology and evolvement (Leopold and Langbein, 1960; Ikeda et al., 1981; Howard, 1984; Braudrick et al., 2011). Since a universal theory of meander morphodynamics remains elusive, the field investigations (Hickin and Nanson, 1984; Dietrich, 1983; Nicoll and Hickin, 2010), theoretical analyses (Ikeda et al., 1981; Blondeaux and Seminara, 1985; Parker et al., 1982), and numerical simulations (Sun et al., 1996; Lancaster and Bras, 2002; Parker et al., 2011) have not provided a complete insight into the processes that govern the meander morphology. The planform geometry characteristics of meanders and oxbow lakes in floodplains are still not completely understood, especially since sufficient field data is still lacking. A meandering river exhibits progressive change in position as it migrates across its floodplain.

In the past 20 years, the geometric features of meandering rivers and oxbow lakes morphology have drawn interests from many river scientists. Oxbow lakes are known to form along meandering rivers through neck cutoffs, which can be surveyed by satellite images. Using satellite images from Google Earth, Constantine and Dune (2008) measured the meandering channels and oxbow lakes characteristics of 30 large meandering rivers to identify the controls of the production of oxbow lakes by neck cutoffs. Zhang et al. (2008) studied the characteristics of incised meanders in the Jialing River of China based on images from Google Earth as well. Nicoll and Hickin (2010) did similar work to investigate the planform geometry and channel migration of confined meandering rivers on the Canadian prairies. Guneralp et al. (2009) analyzed the relation between the plane curvature and the evolution rate of meandering rivers using the method of discrete signal process. Gunderalp's study determined a highly

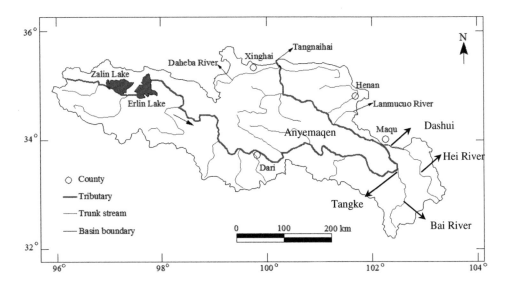

Figure 3.1 Yellow River source and locations of the Bai and Hei rivers.

nonlinear relation between the plane curvature and the evolution rate of meandering rivers and how this relation changes with the river position.

The authors of this book observed many meandering rivers in the Yellow River source during field surveys in 2010–2014. It is impossible to study all meandering rivers in this area, but it is feasible to choose several typical meandering rivers as examples, such as the Bai and Hei rivers and the Lanmucuo River. The Bai and Hei rivers in the Ruoergai Basin are tributaries of the Yellow River, of which the morphological characteristics were studied using Google Earth satellite images and digital elevation model (DEM) data.

3.1.1 Meandering tributaries of the Yellow River

The Yellow River flows out of the Zhaling Lake and Erlin Lake along the northern valley terrain of the Bayanhar Mountain to the southeastern direction, traveling through the valley reach in Jiuzhi County and flowing through the floodplain in Maqu and Ruoergai counties. The Yellow River source encompasses the eastern edge of the Anyemaqen Mountain from the southeast to northwest direction in the Ruoergai Basin, which changes its direction by nearly 180 degree and forms a great bend of 'U' shape, which is called the Yellow River First Bend (Fig. 3.1). The bend is 270 km long. The entry location of the bend is in Awancang Town of Maqu County, and the bend apex is located in Tangke Town of Hongyuan County. The lower end of the bend is located at the southwest side of Maqu County.

The Gannan grassland is on the left side of the Yellow River First bend, where several small meandering rivers flow into the Yellow River. The right side of the bend is the Ruoergai wetlands where the Bai River flows into the Yellow River at Tangke

Table 3.1 Features of the Bai River and Hei River.

Name	Channel length (km)	Basin area (km²)	Annual discharge (10⁶ m³)	Annual suspended load (10³ t)	Suspended load concentration (kg m⁻³)
Bai	270	5488	2070	429	0.21
Hei	456	7068	1030	342	0.33

Town and the Hei River flows into the Yellow River at the southeast of Maqu County (Fig. 3.1).

The Bai and Hei rivers originate in Sichuan Province and the northeastern region of the Plateau. The two river basins are located at 32°20′– 34°16′ N, 102°11′ – 103°30′ E. Good vegetation cover, wetlands, grasslands, and dense bush have developed in the two basins in their source region, floodplain, and hills on the middle stream. The Bai and Hei rivers are naturally-developed meandering rivers, which have experienced little impact from human activities and have well-developed vegetation and highly meandering channels. Table 3.1 shows the basin features at the Tangke and Dashui hydrological stations (Fig. 3.1) on the Bai and Hei rivers, respectively (Zhao, 2005).

The Bai River originates in the eastern end of the Bayanhar Mountain. It has a source altitude of 4460 m with weak peat swamp along the river, flowing through Hongyuan County from south to north, and entering the Yellow River at the north of Tangke Town. The total length of the Bai River is 270 km long with a drainage area of 5488 km² and an average longitudinal gradient of 0.55‰. At the Tangke hydrological station on the Bai River, the annual average discharge is 65.6 m³/s, the annual runoff is about 2.07×10^9 m³, the annual mean suspended load is 429×10^3 t, and the annual mean sediment concentration is 0.21 kg/m³ (Table 3.1). The Bai River branches into two anabranches at the mouth of the confluence into the Yellow River. It has a large number of oxbow lakes and abandoned river channels along its course, and its tributaries are mostly meandering rivers. The upper reach of the Bai River is confined by the hills. The river morphology of the Bai River is characterized by a non-uniform distribution of asymmetric dendritic tributaries. There are many well-developed tributaries at the river source, but only scarce tributaries in the downstream reach. The Bai River begins with narrow river valleys at the mouth, and then widens as it flows into the Ruoergai grassland while the sinuosity increases (Fig. 3.2a).

The Hei River originates in the western Min Mountain. It has an altitude of approximately 4335 m, flowing through Ruoergai County from northwest to southeast, meandering through marsh grass, entering the Yellow River at southeast of Maqu County. The Hei River is about 456 km long with a drainage area of 7600 km² and an average longitudinal gradient of about 0.16‰. At the Dashui hydrological station on the Hei River, the annual average discharge is 32.6 m³/s, the annual runoff is 1.03×10^9 m³, the annual mean suspended load is 342×10^3 t, and the annual mean sediment concentration is 0.33 kg/m³ (Table 3.1). The water elevation where the Bai River enters the Yellow River is 3416 m. It has a series of typical meandering channels with a large number of oxbow lakes and abandoned channels along the river. Confined

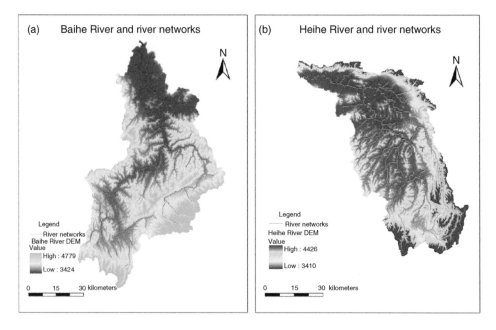

Figure 3.2   DEM and river networks morphology of Bai River (a) and Hei River (b).

by the hills along the upper reaches of the Hei River, the tributaries are asymmetric, and the number of tributaries decreases along the river from upstream to downstream (Fig. 3.2b).

3.1.2   Sinuosity of the Bai and Hei Rivers

The longitudinal profile shows the change of water surface elevation along the rivers and reflects the topography and the hydraulic geometry of the drainage basin. The starting points of the actual longitudinal profiles of the Bai and Hei rivers are clear and the channel morphology is easy to identify. The starting point of the Bai River is Longriba Town (32°26′41″N, 102°21′27″E) in Hongyuan County's, and the starting point of the Hei River is Rangli Village (33°02′18″N, 103°00′38″E) in Ruoergai County.

The longitudinal gradient and the channel width of the Bai and Hei rivers have always been studied together because both parameters are the primary characteristics of a river channel. The Bai River has an average longitudinal gradient of 0.55‰. The gradient has a maximum value of 0.68‰ at the source region, then stays flat for a short distance, and finally gradually decreases to 0.42‰. The Hei River has an average longitudinal gradient of 0.16‰. The gradient has a maximum value of 0.34‰ at the source region, then stays flat for a short distance, and then gradually decreases to 0.07‰. It is obvious that the longitudinal gradient and channel width of the Bai River are greater than those of the Hei River. The trend of the river widening is also more evident with the Bai River (Fig. 3.3).

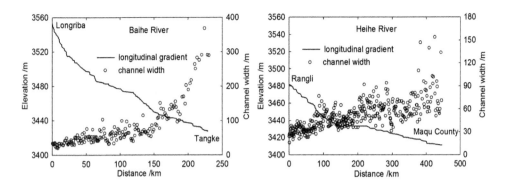

Figure 3.3 Longitudinal profiles and channel widths along the courses of Bai and Hei rivers.

The channel width of the Bai and Hei rivers gradually increases from upstream to downstream. The average channel width of the Bai River is about 77 m, and the average channel width of the Hei River is only about 52 m. The Bai River widens faster and larger than the Hei River. Although located in similar floodplains, the characteristics of the two rivers are directly affected by the annual average discharge, which is 65.6 m³/s for the Bai River, and 32.6 m³/s for the Hei River.

3.1.2.1 Extremely high sinuosity

Sinuosity is the most important parameter for characterizing a meandering river, generally defined as the ratio of the centerline length of the channel to the centerline length of the valley (Stølum, 1998; Hooke, 2007; Zhang et al., 2008). The sinuosity of meandering rivers is conveniently measured by a dimensionless parameter, $S = L/d$ where S is denoted as the sinuosity, L is the length of the river along its centerline between two points and d is the straight line distance between the two points. If the river channel is nearly straight, the sinuosity has a minimum value of 1. The sinuosity of a meander reaches the maximum value where a neck cutoff occurs. The sinuosity depicts the curvature and morphology complexity of meandering rivers.

In order to reflect the average bending degree of a meandering channel, the sinuosity is typically measured for a series of continuous meanders instead of a single meander. Since the meander size of the Bai and Hei rivers is different, the channel length of the continuous meanders measured for calculating the sinuosity was not the same. In order to ensure the integrity and consistency in the computation, the channels measured varied between 1.2 km and 7.8 km (Fig. 3.4).

The sinuosity of the Bai and Hei rivers reflects the curvature morphology of the local river reaches. The sinuosity of the Bai and Hei rivers along the courses decreases. The maximum sinuosity of is about 2.78 for the Bai River and about 5.47 for the Hei River (Fig. 3.4). In general, very meandering stream has maximum sinuosity about 2. The sinuosity larger than 5 is very unusual. The average sinuosity is 2.46 for the Hei River and 1.68 for the Bai River, both being greater than 1.3. The average sinuosity of the Hei River is significantly greater than that of the Bai River. Some reaches of the

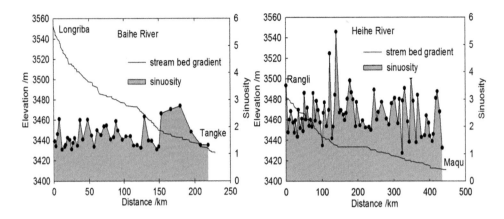

Figure 3.4 Stream bed gradient and sinuosity along the Bai and Hei rivers.

Hei River has much higher sinuosity, which directly reflects the complexity and curvature of its morphology.

The river basins of the Bai and Hei rivers are both located in the northeastern region of the Plateau. They have very similar climate, geology, topography, and vegetation. The sinuosity of the Hei River is, however, significantly greater than that of the Bai River. There are two reasons for this difference. Firstly, the flow discharge in the two rivers, which is the dominant force for shaping the channel morphology, is different. The average annual flow discharge for the Bai River is $65.6 \, m^3/s$, while the average annual flow discharge for the Hei River is only $32.6 \, m^3/s$. The high flow discharge in the Bai River strengthens the flow's capability of scouring the riverbank and widening the channel, which is not conducive to the continuous development of meanders. Meanwhile, the lower discharge in the Hei River adjusts to the continuous development of meanders which results in higher sinuosities. Secondly, the average channel longitudinal gradients of the Bai River and Hei River are 0.55‰ and 0.16‰, respectively (Fig. 3.4). The development of a meandering river needs to decrease to a low longitudinal gradient in order to extend the flow path. In addition, the high frequency of cutoffs along the Hei River results in a large number of oxbow lakes. The high sinuosity and ample oxbow lakes along the Hei River reflect the low channel gradient.

3.1.2.2   Relation between channel width and sinuosity

The relation between the channel width and sinuosity is not straightforward, but some correlation exists. The channel width of the Bai River and Hei River increases along their course. The trend of the sinuosity change is uncertain. Nevertheless, the channel width consistently increases and the river channel widens from upstream to downstream. Figure 3.5 indicates that the channel width and sinuosity have a relation to some extent, i.e. the smaller the channel width is, the higher the sinuosity. This relation is obvious in the middle reaches of the rivers.

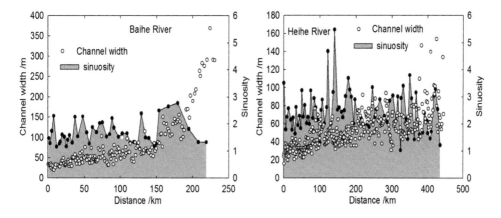

Figure 3.5 Relation between channel width and sinuosity along Bai and Hei rivers.

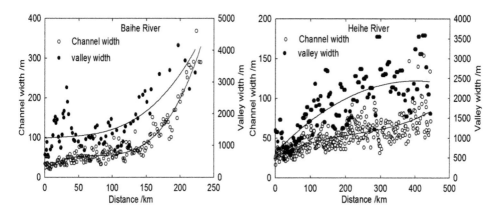

Figure 3.6 Channel width and valley width along Bai and Hei rivers.

3.1.2.3   *Relation between channel width and valley width*

The valley width constrained by the hills reflects the largest distance of the lateral migration of the meandering river. The changes of the valley width and channel width along the river course are similar. The valley width of the Bai River increases rapidly along the first ~31 km of the upstream reach, and then quickly drops to ~1400 m, and then increases again along its course. The Bai River has a valley width in the range of 0.6–4.2 km and a relatively wide channel (Fig. 3.6). The Hei River has a valley width in the range of 0.4–3.6 km and a relative narrow channel.

   The valley width of the Hei River always increases along its course, however, the valley width first increases and then decreases due to restrictions by the hills. Along the downstream reach, the channel width reaches the maximum value, but the valley width reduces quickly (Fig. 3.6). Moreover, the magnitude of the change of the valley width and channel width of the Bai River is greater than that of the Hei River.

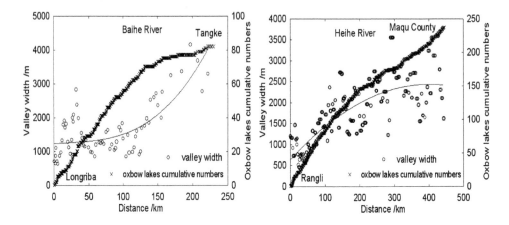

Figure 3.7 Valley width and cumulative number of oxbow lake along Bai and Hei rivers.

The channel width of the Hei River changes slowly. The valley width first increases up to about 2400 m and then decreases. The difference between the valley width and channel width of the Bai River is greater than of the Hei River because the Bai River basin has a wider lateral space.

3.1.3   Oxbow lakes

3.1.3.1   Numerous oxbow lakes

Neck cutoff of meandering rivers produces oxbow lakes, which plays an important role on river geometry and nonlinear characteristics, and reduces sinuosity and complexity. The relation between the oxbow lake production and valley width was investigated by measuring the valley width and recording the number of oxbow lakes along the Bai and Hei rivers.

The numbers of oxbow lakes reflect the times of neck cutoffs, and indirectly indicate the rate and intensity of the lateral evolution of the meanders. The channel length of the Bai River is 228 km. The cumulative number of oxbow lakes along the Bai River is 82, averaging 0.36 per km. The channel length of the Hei River is 445 km. The cumulative number of oxbow lakes is 237 along the Hei River, averaging 0.53 per km (Fig. 3.7).

The number of oxbow lakes along the upper reach of the Bai River increases as the valley width increases. While the valley width continues to increase in the middle reach, the production rate of oxbow lakes decreases. The number of oxbow lakes along the upper reach of the Hei River increases as the valley width increases. In the middle reach of the Hei River, the valley width increases smoothly, and the number of oxbow lakes continues to increase (Fig. 3.7). The changes of the valley width and the number of oxbow lakes along the Bai and Hei rivers have significant differences. The migration rate of the Bai River decreases along its course, and the frequency of neck cutoffs exhibits the same trend. The migration rate of the Hei River tends to be uniform along the entire river and the frequency of neck cutoffs is high.

3.1.3.2 Cutoff of meanders and formation of oxbow lakes

The long-term evolution of a meandering river depends on the flow-sediment discharge and boundary conditions. Field observations have shown that the key control factors for the morphology and evolution of the Bai and Hei rivers include channel gradient, flow discharge and valley width. In general, the development of a meandering river tends to create a low channel gradient by extending the longitudinal length. Assuming that the alluvial river maintains the sediment transport equilibrium by adjusting the slope and geometry to reach a minimum energy dissipation rate under specific local conditions, the value of US, where U is the cross velocity and S is the channel slope, reaches the minimum via increasing the channel length and reducing the channel gradient. The average gradient of the Bai and Hei Rivers is 0.55‰ and 0.16‰, respectively. Accordingly, the average sinuosity for the Bai and Hei Rivers is 1.68 and 2.46, respectively. Figure 3.8 demonstrates that the morphology of the downstream reach of the Bai River is more regular and symmetric than that of the Hei River, the planform of which shows intense transverse migration, which causes the formation of a large number of oxbow lakes.

The basin area and channel length of the Hei River are much greater than those of the Bai River, however, the average annual flow discharge of the Bai River is approximately 2 times larger than that of the Hei River (Table 3.1). The flow discharge of the Bai River directly increases the river's capability of lateral widening and bed incision, hence, the channel width of the Bai River is obviously greater than that of the Hei River, especially in the downstream reach (Fig. 3.8). Channel widening and incision reduce the average flow velocity, but are not beneficial to the river migration. Hence, the lateral migration rate of the Hei River is much higher than that of the Bai River, which is embodied in the cumulative number of oxbow lakes along the course. The sinuosity and cutoff frequency of the Hei River is evidently higher due to narrower channel width.

As both sides of the Bai and Hei rivers are hills, the valley width controls the maximum lateral distance of the channel migration. To some extent, the two rivers are not completely free winding. The relation between the valley width and channel width is positive, which reflects not only the watershed characteristics, but shows the constraint effect of the valley width. In addition, with smaller valley width in the upper reaches, there are a large number of oxbow lakes along the Bai and Hei rivers. However, in the lower reach of the Bai River, as the valley widens rapidly, the number of oxbow lakes decreases, while the number along the Hei River remains high (Fig. 3.6 and Fig. 3.8).

Currently, the hydrology and cross section geometry data of the Bai and Hei rivers are very limited and hard to obtain. This book does not deeply investigate the quantitative relation between the geometrical morphology and flow energy dissipation. The Bai and Hei rivers entering the Ruoergai grassland have similar climate, geology, and vegetation. The geometrical planform difference between the Bai and Hei rivers is reflected in the density of oxbow lakes (Fig. 3.7) and sinuosity (Fig. 3.9). The density of oxbow lakes per unit river length along the Hei River is distinctly more than that

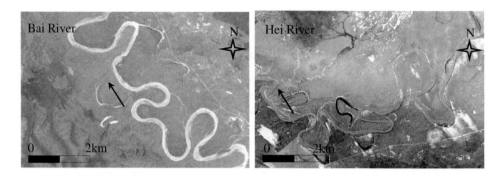

Figure 3.8   Images of downstream reaches of the Bai and Hei rivers from Google Earth (Oct. 2009).

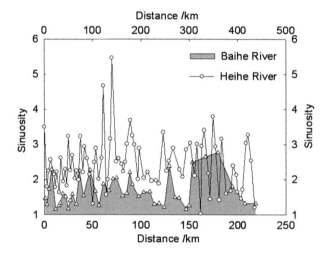

Figure 3.9   Sinuosity along Bai and Hei rivers.

along the Bai River. The corresponding migration rate and cutoff frequency of the Hei River are higher than that of the Bai River. The sinuosity change is 1.85–5.85 for the Hei River and 1.13–2.78 for the Bai River, indicating that the Hei River has higher complexity and migration rate. The Hei River basin has suitable conditions for the development of a meandering river.

3.2   DEVELOPMENT PROCESS OF MEANDERS

In the last twenty years, many researchers have paid attentions to riparian vegetation which resists flow erosion, enhances the stability of riverbanks, reduces nearbank velocity, and plays an important role in restoring river ecology (see Kort et al., 1998; Prosser et al., 1995; Abernathy and Rurtherford, 2001; Hubble et al., 2010). Field investigations during 2011–2014 in the Yellow River source showed that many tributaries of the Yellow River are meandering rivers such as the Zequ, Lanmucuo, Requ, Jiqu, Haqu, Maiqu, and Gequ rivers. These meandering rivers can be easily identified and have extremely high sinuosity. More importantly, plateau meadow grows on the riverbanks, therefore, these rivers are called "meadow meandering river".

The upper layer of the riparian banks in the meadow meandering rivers has strong resistance to erosion, for which riverbank failures are obviously different from that for alluvial rivers in fluvial plains. In meandering rivers in fluvial plains, when the flow scours the non-cohesive soil on the riverbanks to a certain depth, the upper soil body reaches a critical balance, consequently, shear failures occur on the banks along the shear plane. The effect of vegetation is generally negligible in this process. Nevertheless, the meadow layer forming an intense soil-root complex on the riparian banks of meadow meandering rivers has very strong resistance to shear and erosion. Only when the non-cohesion soil in the lower layer is scoured out and the upper soil-root complex is fully hanged like a cantilever, which reaches the critical balance under the action of self-weight, that the tensile failure occurs. To date, cantilever bank failures have been studies by other researchers (Greg and Lance, 2002; Motta et al., 2012), but meadow cantilever bank failures have not been studied.

The concave bank of meadow meandering rivers is almost perpendicular to the bed. Slump blocks of soil-root complex vertically fall to the bank toe, which protects the non-cohesive sediment for several flooding periods and inhibits nearbank erosion. After the slump blocks are soaked in water for a long time, the root complex rots and decomposes, losing the interlocking function between the soil and roots, and finally the blocks are carried away by the nearbank flow. It is necessary for the inhibitory scour effect of the slump blocks to be consistent with the deposition rate of the convex bank in order to maintain an appropriate rate of transverse migration of the meanders (Burckhardt and Todd, 1998; Micheli et al., 2002). Recently, Parker et al. (2011) in their new frame of meandering river simulation model adopted an empirical coefficient to correct the migration rate, which approximately considered the inhibitory scouring effect of slump blocks. Asahi et al. (2013) and Eke et al. (2014) further actualized the physical basis of the coefficient so it could be practically calculated in their numerical models. Therefore, the meadow riparian bank and slump blocks of a cantilever bank failure play important roles in keeping the synchronous process of the concave and convex banks and maintaining continuous meandering channels (Dulal et al., 2010). In order to study the mechanism of bank failures in meadow meandering rivers, field investigations, laboratory tests and theoretical analyses were performed to study the geometry of slump blocks, the mode of the bank failure, critical stress conditions, and the characteristics of the soil-root complex. These studies were carried out to improve the understanding of the uniqueness of the meadow slump blocks and their ability to inhibit lateral evolution of meanders.

3.2.1 Meandering river in alpine meadow

The Lanmucuo River in the Yellow River source is a typical meadow meandering river (Fig. 3.10), originating at 34°26′N, 101°29′E. Field investigations of the Lanmucuo River were conducted four times between 2011 and 2014. The river basin has a

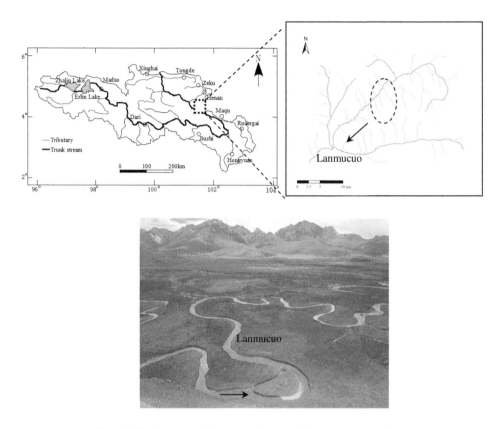

Figure 3.10 Location of Lanmucuo River and successive meanders.

plateau continental climate in the humid subarctic climate zone. It is located at elevation 3400–4200 m with an average annual temperature of −4°C and a multi-year average precipitation of 329–505 mm. The vegetation types within the Lanmucuo basin include only alpine meadow, plateaus-cold meadow and alpine grassland meadow, and cushion and rock-patch sparse vegetation in local high altitude areas.

The field investigations and sampling during 2011–2014 included hydrological parameters (e.g. hydraulic gradient, flow velocity, channel width, and water depth), geotechnical characteristics of the riverbank (i.e. soil grain size, soil moisture content, and bulk density), vegetation composition, and soil-root complex. The samples taken from the concave and convex banks were $50 \times 50 \, \text{cm}^2$ and $100 \times 100 \, \text{m}^2$ in size, respectively. They were used to determine the riparian vegetation composition, vegetation coverage, and species diversity. Eight on-site undisturbed soil-root complex were taken using large plastic pots and 16 samples were taken using ring knives in the concave bank. These samples were taken back to the laboratory, and their density, moisture content, root density, cohesion and internal friction angle were measured and shown in Table 3.2.

Table 3.2 Measured parameters of slump blocks and critical blocks in Lanmucuo River.

Type	Number	Average length (m)	Average width (m)	Average thickness (m)	Main root length (m)
Slump block	63	1.940	0.847	0.845	0.571
Critical block	15	1.562	0.761	0.699	0.416
Average		1.751	0.804	0.772	0.494

3.2.2 Physical characters of slump blocks

Observations made at the field investigations indicated that the concave bank toe of meadow meandering rivers was protected by slump blocks, which protected the bank from scouring. This book classifies slump blocks into collapse slump blocks and critical blocks. The former refers to blocks collapsing from the riverbank and dumping into the water or attaching to the slope. The latter refers to a situation when the riverbank is in a critical condition and cracks along the bank appear, but the blocks are not collapsing and resulting in a complete tensile failure. The thickness of slump blocks depends on the root length and density, but the length of slump blocks are changing along the course. The lateral width of slump blocks depends on the transverse scour width and the tensile strength of soil-root complex. The upper layer of the riverbank in the Lanmucuo River is meadow soil-root complex, the middle layer is sand, and the lower layer is gravel. The geometric data (width, length, and thickness) of the slump blocks are useful in the analysis and calculation of the critical stress condition of a bank failure.

The investigated reach of the Lanmucuo River was about 5 km long. The geometric characteristics and root length of 15 critical blocks and 63 slump blocks as shown in Figure 3.11 and Figure 3.12. As shown in Table 3.2, the width of the slump blocks ranged between 40 and 130 cm, the length ranged between 100 and 630 cm, and the root length ranged between 20 and 120 cm. There is no obvious trend of the relation between the block width and length. The change of the block width and root length, however, shows the same trend, i.e. the greater the root length, the wider the block. Roots play a stronger role in the soil consolidation and winding, therefore, the root length has a significant linear correlation with the block thickness.

Table 3.3 shows the density, moisture content, root density, cohesion and internal friction angle of four typical samples of slump blocks. The density and moisture content values are very close between the four samples with the mean values being 1.63 g/cm³ and 47.36%, respectively. The root densities of the four samples are very different because the vegetation density of the riparian bank is not uniform.

3.2.3 Bank failure in the meandering process

Slump blocks mainly exist in concave banks and have certain protective effect on riverbanks (Fig. 3.13) by inhibiting the lateral evolution rate. Our study found that a majority of the slump blocks formed within two year. Old blocks from the previous year were probably taken away by the flow this year or the next year because of

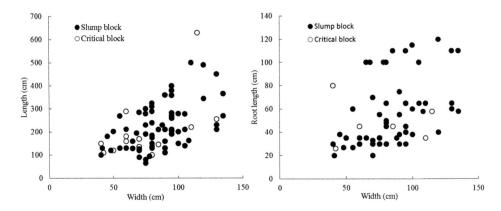

Figure 3.11 Relation between length and width (left) and relation between width and root length of slump blocks (right).

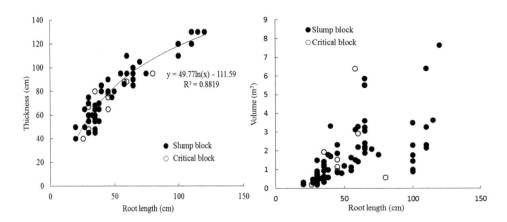

Figure 3.12 Thickness of slump blocks as a function of the root length (left) and volume of slump blocks as a function of root length (right).

Table 3.3 Physical parameters of undisturbed root-soil complex in Lanmucuo River.

No.	Density (kg/m³)	Moisture content (%)	Mean root diameter (×10⁻³ m)	Resistance tensile strength (kPa)	Soil resistance shear strength (kPa)
B-1	1542	53.09	0.93	27680	11.22
B-2	1516	49.88	1.13	26070	12.15
B-3	1559	42.78	1.36	21920	10.70
B-4	1528	43.69	2.11	18930	11.54
Average	1536	47.36	1.38	23650	11.40

Figure 3.13 Cantilever slump blocks and root layer in a concave bank.

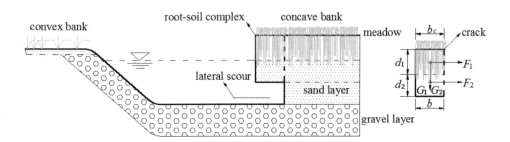

Figure 3.14 Scheme of a slump block.

water erosion, decaying and freeze thawing. This means that the protection time of slump blocks is about 1–2 years. It should be noted that the accurate protection time of slump blocks still needs to be verified by continuous in-situ measurement for a number of years because this area lacks historical remote sensing image with high precision.

The form of a bank failure in meadow meandering rivers under the action of gravity is a cantilever tensile failure in which destructive cracks begin on the surface and cut through the entire soil-root complex layer. During the entire failure process, the shear effect is negligible. Figure 3.13 and 3.14 show the three layers and a slump block unit.

Assuming that the critical block is a cuboid, based on the unit length critical block stress condition, the following moment balance equation applies:

$$(G_1 + G_2) \times \frac{b_c}{2} = F_1 \times \frac{d_1}{2} + F_2 \times \left(d_1 + \frac{d_2}{2}\right) \tag{3.1}$$

where G_1 and G_2 are the masses of the soil-root complex and sand transition layer, respectively. $G_1 = \rho_1 g b_c d_1$, $G_2 = (\rho_2 - \rho_w) g b_c d_2$, where ρ_1, ρ_2, and ρ_w are the densities of the soil-root complex, sand transition layer, and water, respectively, g is the acceleration of gravity, d_1 and d_2 are the thicknesses of the soil-root complex and sand transition layer, and b_c is the critical transverse width. F_1 and F_2 are the maximum resistance shear strength of the soil-root complex and the maximum cohesive force of the sand transition layer, respectively. $F_1 = (S_0 + \Delta S_1) d_1$, where S_0 is maximum resistance shear strength of soil unit area, and ΔS_1 is incremental resistance shear strength induced by root system. $F_2 = c_2 d_2$, where c_2 is the cohesion of sand layer. It is assumed that the root system crossing over collapsing face all reaches limit resistance tensile strength and occurs broken simultaneously. Therefore, Wu et al. (1979) derived a root incremental model to give an approximate equation: $\Delta S_1 = 1.2 T_N (A_r/A_s)$, where T_N is the mean resistance tensile strength of the root system per unit area, A_r/A_s is the ratio of the root sectional area to the collapsing face area, and $A_s = d_1 b_L$.

Inserting the expression of G_1, G_2, F_1, and F_2 into Eq. (3.1), the following equation is obtained:

$$\frac{b_c}{2}[\rho_1 g b_c d_1 + (\rho_2 - \rho_w) g b_c d_2] = [S_0 + 1.2 T_N (A_r/A_s)]\frac{d_1^2}{2} + c_2 d_2\left(d_1 + \frac{d_2}{2}\right) \quad (3.2)$$

Simplifying Eq. (3.2) to obtain the critical tensile strength, Eq. (3.3) derived as below:

$$F_1 = \frac{g b_c^2 [\rho_1 d_1 + (\rho_2 - \rho_w) d_2] - c_2 d_2 (2d_1 + d_2)}{d_1^2} \quad (3.3)$$

Analyzing the variables of the tension failure, a critical balance function of the tensile failure is derived as follows:

$$f = (G_1 + G_2) \times \frac{b_c}{2} - F_1 \times \frac{d_1}{2} - F_2 \times \left(d_1 + \frac{d_2}{2}\right) \quad (3.4)$$

$$f = \frac{b_c}{2}[\rho_1 g b_c d_1 + (\rho_2 - \rho_w) g b_c d_2] - [S_0 + 1.2 T_N (A_r/A_s)]\frac{d_1^2}{2} -$$
$$c_2 d_2 \left(d_1 + \frac{d_2}{2}\right) \quad (3.5)$$

Eq. (3.5) can be simplified as follows:

$$f = \frac{g b_c^2}{2}[\rho_1 d_1 + (\rho_2 - \rho_w) d_2] - \frac{d_1^2}{2}[S_0 + 1.2 T_N (A_r/A_s)] - \frac{c_2 d_2}{2}(2d_1 + d_2) \quad (3.6)$$

where g, ρ_1, ρ_2, and ρ_w are roughly constants in the same river reach, S_0, T_N, A_r, and c_2 are measured by sampling. And d_1, d_2, and b_c are also measured by sampling. The value of d_2 depends on the flow scouring height, when scouring occurs near the bottom of the soil-root complex, i.e., $d_2 \to 0$. Eq. (3.6) can be simplified as follows:

$$f = \frac{1}{2}\rho_1 g b_c^2 d_1 - \frac{d_1^2}{2}[S_0 + 1.2 T_N (A_r/A_s)] \quad (3.7)$$

Table 3.4 Parameters of root tensile strength in collapse surface.

d_i (m)	$\bar{d}$ (m)	N_i
≥ 0.002	0.00250	$750 \times 0.36 = 270$
0.0015–0.0020	0.00175	$750 \times 0.184 = 138$
0.0010–0.0015	0.00125	$750 \times 0.137 = 103$
0.0005–0.0010	0.00075	$750 \times 0.319 = 239$

If the slump block reaches the critical equilibrium condition, $f = 0$, therefore,

$$b_c = \left(\frac{d_1[S_0 + 1.2T_N(A_r/A_s)]}{\rho_1 g} \times 100 \right)^{0.5} \tag{3.8}$$

Eq. (3.8) actually neglects the coarse sand and gravel layer below the soil-root complex because this layer is very thin when the bank failure occurs. The parameters of the sixty-three slump blocks and 15 critical blocks are as follows: the thickness of the soil-root complex was $d_1 = 0.772$ m and the density was 1536 kg/m³.

Table 3.4 shows the mean root diameter and average resistance tensile strength of a single root. T_N is given using average resistance tensile strength of four samples, $T_N = 23650$ kPa, $S_0 = 11.4$ kPa. The sum of the sectional area is obtained:

$$A_r = \sum_{i=1}^{n} \frac{N_i \pi \bar{d}_i^2}{4} \tag{3.9}$$

where $\bar{d}_i$ is mean root diameter of the i-th grade root, N_i is number for the i-th grade root. According to the situ measurement, the root diameter is divided into four grades. The total number of each grade is equal to the weighted number of each grade multiplying the total root number (Table 3.4). The average length of slump block b_L was 1.751 m and the total number of roots was $N \approx 750$. Carex Moorcroftii a is unique dominate vegetation, so the resistance tensile of each grade root is calculated in Table 3.4.

Incorporating Eq. (3.9) into Eq. (3.8), and the critical width of the slump block b_c is obtained:

$$b_c = \left(\frac{d_1\left[S_0 + 1.2T_N\left(\sum_{i=1}^{n} \frac{N_i \pi \bar{d}_i^2}{4} / d_1 b_L \right) \right]}{\rho_1 g} \times 100 \right)^{0.5} \tag{3.10}$$

Using Eq. (3.10), the average critical width b_c is 0.512 m, compared with $b_c = 0.804$ m (see Table 3.2), relative error of 16.7%. Eq. (3.10) does not consider the inter-twisting of roots, so the calculated value is lower than the measurement. Nevertheless,

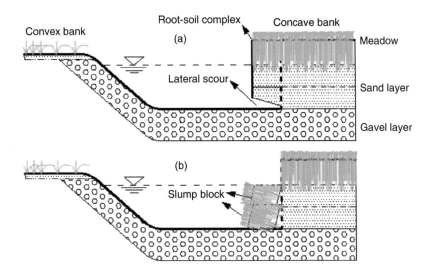

Figure 3.15 Bank scour and slump blocks protection with the root-soil complex.

Eq. (3.10) without any empirical coefficient, is of importance for analyzing the lateral migration rate and protection time of slump blocks.

3.2.4 Transverse migration rate of meanders

A cantilever bank failure occurs when the non-cohesive sediment (sand and gravel) of the lower layer of the riverbank is scoured away in the absence of the slump block protection, and the flow further scours until the riverbank reaches a critical equilibrium state and longitudinal cracks appear along the course. Meanwhile, the upper blocks lose their support because the flow scours the sand beneath the critical block. The two processes, non-cohesive layer scouring and slump blocks toppling the riverbank (Fig. 3.15), are very important for the understanding of the transverse evolution rate of meanders.

At $t = t_0$, it is assumed that the annual flood season just begins and the lateral erosion starts to occur due to the absence of slump block protection. At $t = t_1$, a certain lateral width of the non-cohesive layer has become hollow and the equilibrium condition of the upper soil-root complex is broken and tensile cracks appear. The erosion time can be calculated using the following equation:

$$\Delta t_e = t_1 - t_0 = \frac{b_c}{\zeta_e} \tag{3.11}$$

where ζ_e is the erosion rate of the non-cohesive layer, which is a function of the near-bank flow stress. There is a lack of reliable data and calculation formula to determine ζ_e. At this time, b_c can be regarded as the distance of the meander's lateral movement. The overturning slump block continues to protect the riverbank, and the deposition rate of the convex bank is less than the erosion rate of the concave bank, therefore, the lateral migration of the meander cannot complete. With the slump block soaked into the water, the clay dissolving into the water, the root rotting, and the network structure of the interlaced root dispersed, the weight of the slump block becomes lighter and the block is eventually taken away by water flow.

At $t = t_2$, namely when the protection time is equal to $\Delta t = t_2 - t_1$, the slump block decomposes and completely loses its protection function. The average transverse evolution rate is calculated as follows:

$$\zeta_E = \frac{b_c}{\Delta t_e + \Delta t} \tag{3.12}$$

where b_c can be predicted by Eq. (3.10).

3.3 CUTOFF OF MEANDERS

3.3.1 Types of cutoff

Meandering rivers with broad distribution and aesthetic landscape have attracted many scholars in geomorphology, fluid mechanics, river dynamics, and hydraulic engineering to make unremitting efforts in the past 100 years. The bending instinct of natural rivers had been observed in experimental flume and field surveys for a long time. There are different interpretations on the intrinsic mechanism of river meandering (Da Silva, 2006). Meandering rivers tend to move to the direction of increasing sinuosity, which is the dominate behavior of the upstream and downstream migration (Zolezzi and Seminara, 2001). Hereby meandering rivers display the circuitous processes and self-similar landscape in the floodplain. Researchers have found out that the planform winding migration of meandering river is not unlimited, because natural cutoff is a physical mechanism to control the morphological complexity and it sharply decreases the sinuosity (Camporeale et al., 2008).

Meander cutoff can be generally classified into two types, i.e. neck cutoff and chute cutoff, which have different physical processes and mechanics (Erskine et al., 1982). Indeed, Lewis and Lewin (1983) pointed out that neck cutoff occurs where the up-channel and down-channel limbs of a river bend are less than a channel width apart at the time of breaching, and chute cutoff occurs where a much longer breaching channel is created by floods.

If a bend neck undergoes a long-term evolution and shrinks to a certain critical width, neck cutoff may occur at any time. After a sudden huge flood generates overland flow and quickly erodes the neck or the bank with an extremely narrow neck, it opens up a new short channel while the entrance of the former bend is blocked by deposition until an oxbow lake forms. Neck cutoff has not been successfully simulated in laboratory flumes so far, but a simplified linear model (i.e. the rate of bank erosion is linearly related to the excess bank velocity) can be used to simulate the long-term evolution of neck cutoffs (Howard, 1992; Stølum, 1996).

As another type of meander cutoff, chute cutoff normally has a lower sinuosity than neck cutoff. Chute cutoff needs a high stage flood to scour some locations of the river bend, but the main flow may still flow along the former bend channel, coexisting with the old and new channels for a long time and may not necessarily form an oxbow lake in a short time period. The condition of a chute cutoff can be easily achieved in meandering rivers, therefore, chute cutoff is more common than neck cutoff. In a laboratory flume with key auxiliary materials including clay soil, alfalfa sprouts, and silica flour to enhance the bank strength, chute cutoffs were successfully simulated with partial cutoff characteristics (Ying, 1965; Braudrick et al., 2009; Dijk et al., 2012).

In an effort to facilitate navigation and reduce flood threat, a series of artificial cutoff projects had been performed along the lower Mississippi River and the middle Yangtze River (Smith and Winkley, 1996; Pan et al., 1978). Artificial cutoffs can open up a new channel to shorten the original meander length, but artificial cutoffs are an engineering design problem (Xiong et al., 2002). The morphology and evolution of meandering rivers were originally studied via field observations and measurements. One of the most famous study was performed by Louis Fargue, a famous French engineer. According to his long-term observation of the Garonne River, Fargue summarized six laws which guided early navigation regulations (Hager, 2003). Based on the theory of minimum variance, Langbein and Leopold (1966) advanced the sine-generated curve by approximately delineating symmetrical meanders. Kinoshita (1961) created the Kinoshita-generated curve by adding the skewness and flatness coefficients into a sine-generated curve to depict irregular meanders. Not only meander cutoffs are very easy to observe on the earth surface using remote sensing images, but also similar cutoff events can be found in submarine channels induced by turbidity currents (Lonsdale and Hollister, 1979) and the relict meanders on Mars reported by Howard (2009).

It is very difficult to observe a cutoff event because the process of a cutoff occurrence is relative rapid during the flooding period. Geomorphologists have done a lot of work to understand the time and processes of ancient cutoffs (Handy, 1972; Brooks and Medioli, 2003). Erskine et al. (1992) argued that four alluvial cutoffs along the lower Hunter River could be regarded as indicators of old river channels and floods according to detailed historical records and stratigraphic data. Gay et al. (1998) found fifteen neck cutoffs in the Powder River, most of which were caused by ice jams which had formed in the overflowing water and caused headward-cut of the meander neck. Hooke (1995, 2004) studied the neck cutoffs along the Bollin River for the past 20 years, which all occurred during extreme flood events, resulting quickly widening of the new channel and the formation of many sand bars. Moreover, Hooke (1995, 2004) concluded that neck cutoff is an inherent chaotic behavior. Micheli and Larsen (2011) utilized field surveys and aerial images of the Sacramento River during 1904–1997 in a 160-km-long river reach to investigate 27 full cutoffs and 11 partial cutoffs processes. Thompson (2003) described an unsuccessful channel relocation of the Blackledge River in the late 1950s, which resulted in a large meander cutoff and the abandonment of a portion of the relocated channel. It is evident from the studies above that comprehending the processes and mechanism of meander cutoffs can facilitate understanding of the historical condition, current state and future evolution trend of meandering rivers.

In recent years, field surveys and flume experiments of chute cutoff have produced many important results which provided significant references for neck cutoffs. Constantine et al. (2010) elucidated three different mechanisms of chute cutoffs including the enlargement of swales, headcut extension during locally induced flooding and by the downstream extension of an embayment during a sequence of floods. After the chute cutoff completes, the dynamic processes of the chute channel directly affect the meander migration and sediment transport (Fuller et al., 2003; Grenfell et al., 2011;

Ghinassi, 2011). Zinger et al. (2011) observed the occurrence of two chute cutoffs at the Mackey Meander of the Wabash River via remote sensing images and field surveys in June 2008 and June 2009. The two cutoff events triggered strong erosions rapidly, sluicing the sediment into the downstream channel, where the migration rate was one to five times larger than that for a normal lateral migration of the local river bend.

Recently, two flume experiments of chute cutoffs were completed by Braudrick et al. (2009) and Dijk et al. (2012), which successfully simulated the occurrence, processes and conditions. The former used the alfalfa sprouts to increase the bank strength and the latter added silt-sized silica flour to simulate the sediment feed to encourage floodplain formation. The two experiments above may provide a reference for simulating the occurrence of neck cutoffs in the flume, assuming that there is enough experimental run time for the formation of a meander neck.

Neck cutoffs generally occur during high floods. The cutoff processes have rather short durations, therefore, it is not easy to observe these processes directly. In contrast, numerical simulation of meandering rivers can facilely mimic the processes of neck cutoffs (Howard, 1992; Stølum, 1996; Sun et al., 1996, 2001; Stølum and Friend, 1997). The meander morphodynamics models bring some new knowledge about cutoff mechanism (Camporeale et al., 2005), which shows that cutoffs contain a dual role including constraining morphological complexity and producing an intermittent noise which influences the spatiotemporal dynamics (Camporeale et al., 2008). Although a great number of mathematical models on the meandering morphodynamics have been developed (Ikeda et al., 1981; Howard, 1992; Sun et al., 1996, 2001; Duan and Julien, 2010), new physical-based models have emerged quickly in the mean time (Pittaluga et al., 2009; Parker et al., 2011; Motta et al., 2012; Posner and Duan, 2012; Eke et al., 2014). Nevertheless, previous numerical simulations assumed that the channel width remained constant and the erosion rate of the concave bank was equal to the siltation of the convex bank. A critical problem exists in this approach which assumes that the neck cutoff occurs when the neck width is equal to 1.5 times the original channel width, and then the new centerline replaces the original centerline, starting a new round of iteration until the next neck cutoff occurs (Howard, 1992; Stølum, 1997). This assumption is not fully consistent with natural neck cutoffs. Actually, a neck cutoff is induced by a neck bank failure and high flood erosion, but the neck width remains much wider than the channel width.

Field observations and flume experiments have mostly focused on chute cutoffs, while observations and analyses have not been conducted to fully understand the mechanism of neck cutoffs. In order to achieve better physical modeling of the long-term evolution of meandering rivers, it is essential to study the neck cutoff phenomena, processes and mechanism. The studies carried out by the authors of this book utilized a large number of remote sensing images to study the neck cutoffs that were in the process of occurring or near completion.

3.3.2 Neck cutoff

Neck cutoff is a mutation event in the morphodynamic processes of meandering rivers, which not only quickly reduces the sinuosity and complexity of subsequent meanders, but also causes the new channel to initiate a new round of evolution. Neck cutoff has been regarded as a self-organized critical event (Stølum, 1996), an intermittent

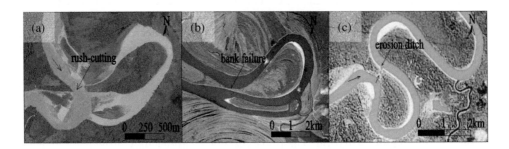

Figure 3.16 Three patterns of neck cutoff (Google Earth images).

noise (Camporeale et al., 2008), and an inevitable event (Gay et al., 1998). These different understandings manifest that the mechanism of neck cutoff has not reached consensus. The sinuosity of the meander planform increases over time, transversely migrating and creeping to extend the longitudinal length, and gradually narrowing the neck width. The shortest distance on the inner floodplain from the upstream inlet to the downstream outlet is referred to as the bend neck.

At present, the types of chute cutoff have been reported by Constantine et al. (2010), but the types of neck cutoff have not been reported. Neck cutoffs are classified into three types: rush-cutting, bank failure, and erosion channel based on the planform, conditions and processes (Fig. 3.16). The rush-cutting type neck cutoff occurs when the overbank flow follows the maximum gradient in an extremely high flood: the flood scours and intercepts the bend neck by cutting through the neck and rapidly forming a new channel. The entrance of the original meandering channel is gradually blocked by deposits and eventually becomes an oxbow lake (Fig. 3.16a). The conditions of rush-cutting are a narrow width of the neck and an extreme flood event which is the most fundamental driving force. Rush-cutting most commonly occurs in the middle Yangtze River, Hanjiang River, Amur River, Hei River, Bai River, Amazon River basin, and Siberia alluvial rivers.

The bank failure type neck cutoff occurs when the sinuosity approaches a critical state and the channel forms an Ω-shape. A large number of neck cutoffs shown on remote sensing images have neck widths of 1/3 to 1/8 of the average channel width, but surprisingly these neck cutoffs are not caused by floods (Fig. 3.16b). Actually, the neck bank is scoured by the nearbank flow, where erosion of the bank toe results in an ongoing bank failure until the neck naturally interconnects. Neck cutoffs of the bank failure type in the Amazon River and Irtysh River have similar bank materials and abundant runoff but do not necessarily need extreme flood events. The natural interconnection of meander necks induced by bank failures is common to find in the Amazon River basin, alluvial rivers of Siberia, and downstream of the Wei River (Pang, 1996).

The erosion channel type neck cutoff occurs if the overbank flow with relative high level scours the floodplain of the inner neck and forms one or several embryonic channels. Because there are relatively wide necks and vegetation protection, a single flood event is insufficient to completely downcut the neck to form a new river channel (Fig. 3.16c). After the embryonic channel formed, even if the water level of the next flood is lower than the previous flood, it can still scour through the channel again, and a new channel will likely form after several times of erosion. The conditions of the erosion channel type neck cutoff are (1) the meander is quite curved with a certain neck width, which is roughly equal to the average channel width; and (2) the overbank floods scour the neck several times until the new channel forms. There are several similarities between the erosion channel type and chute cutoff, but the fundamental difference is that the erosion channel type neck cutoff occurs in the neck location and the chute cutoff occurs without forming the narrow neck. The erosion channel type neck cutoff is a result of several times of floods erosion, which is common in the Amazon River basin, alluvial rivers in North America, and downstream of the Weihe River.

3.3.3 Mechanism of cutoff

3.3.3.1 Rush-cutting type

Rush-cutting is the most prevalent phenomena of neck cutoffs. It is closely related to the frequency, stage and discharge of extreme floods. The direct driving force of the rush-cutting is the strong erosion on the inner neck by the flood, which opens a new channel because the overbank flood seeks the maximum gradient path. It is not critical to consider the headcut erosion because the neck width is very narrow and the scour rate is very high. Therefore, the process of the rush-cutting consists of two steps: (1) the overbank flow erodes the neck surface and transports the vegetation and sediment; and (2) the new channel deepens and widens until it cuts through the upstream and downstream water. Five rush-cutting cutoff events have occurred in the middle Yangtze River since about 130 years ago (Pan, 1999), including the Guchangdi cutoff in 1887, the Chibakou cutoff in 1909, the Hekou cutoff in 1911, the Nianziwan cutoff in 1949 and the Shatanzi cutoff in 1972. Unfortunately, only the cutoff positions were known for these events, and little observations were made during the events. The latest neck cutoff of the middle Yangtze River completed in Shatanzi on July 22–31th, 1972, however, the cutoff development rate was rapid and on-site data was not available.

This section applies the scour rate formula by Wang et al. (2001) which was derived based on a series of flume experiments on high strength scouring processes. The simplification and assumptions are given as follows: (1) the infiltration pressure and groundwater effect do not consider the existence of vegetation; (2) the sediment cohesion is not considered in order to maintain the applicable scope of the scouring rate formula; and (3) the new channel width is far greater than the scouring depth.

During the first step of the rush-cutting, the overbank flow scours the meander neck. The scour rate formula by Wang et al. (2001) is written as follows:

$$S_r = \eta \frac{\gamma}{\gamma_s - \gamma} \frac{J^{1/2}}{d^{1/4}} \left[\frac{\gamma Q_2 J}{B_c} - \frac{1}{10} \frac{\gamma}{g} \left(\frac{\gamma_s - \gamma}{\gamma} gd \right)^{3/2} \right] \tag{3.13}$$

where η is the experimental scouring coefficient, $\eta = 0.218$, which was calibrated by Wang et al. (2001); S_r is the surface erosion rate of the overbank flow; γ' is the dry sediment bulk density; J is the overbank flow gradient, $J = h/(b + B)$; d is the median particle diameter; Q_2 is the overbank flow discharge in the new channel; and B_2 is the trench bottom width after the neck surface scouring.

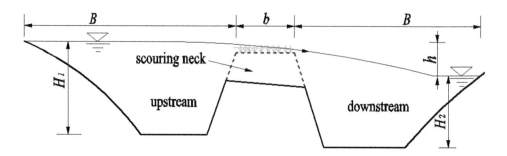

Figure 3.17 Schematic diagram of rush-cutting pattern.

During the second step, the deep groove forms, the depth of which gradually increases from zero to a certain value as the cutoff process accelerates. The change of the scouring depth with time can be described as follows:

$$\frac{dH_c}{dt} = \frac{S_r}{\gamma_s \rho_0 B_c} \tag{3.14}$$

where H_c is the scouring depth, t is time, and ρ_0 is the porosity.

The cutoff time is very short due to the fact that rush-cutting type neck cutoffs occur in a high-level flood period. It is essential to divide the neck scouring into three stages. At the first stage, the overbank flow scours early and the flow width is much larger than the water depth. With the formation of the new channel, the overbank flow downcuts into the new channel. During this stage, the flow discharge in the new channel is assumed to remain constant, and the new channel is too shallow to collapse and its bottom width stays constant for a period of time. At the second stage, concentrated erosion occurs in the new channel, and it can be assumed that the scouring rates of the depth and width change in a linear relation until the scouring depth reaches half the channel width. Simultaneously, the discharge entering the new channel increases, which can be expressed as follows:

$$\frac{dH_c}{dt} = \frac{dB_c}{dt} \tag{3.15}$$

At the third phase, the neck is continuously eroded; hence, the new channel may have a certain degree of siltation. If the flow reaches the transport capacity of water and sediment, the new channel will achieve a temporari stability and no longer broadens or deepens. Since no field data is available, the above formula and described processes cannot be directly verified.

The first stage is the formation of a new channel created by overbank flow. The erosion rate of this process can be simply estimated using Eq. (3.14). During the second stage, the discharge Q_2 increases as the downstream water depth decreases, hence, the new channel slope increases. Eq. (3.14) and Eq. (3.15) can be integrated into the following form:

$$H_c = 0.218 \frac{\gamma}{\gamma_s (\gamma_s - \gamma)} \frac{J^{1/2} t}{\rho_0 B_c d^{1/4}} \left[\frac{\gamma Q_2 J}{B_c} - \frac{1}{10} \frac{\gamma}{g} \left(\frac{\gamma_s - \gamma}{\gamma} g d \right)^{3/2} \right]$$ (3.16)

Eq. (3.16) shows that the scour depth H is proportional to Q_2 and J, and inversely proportional to B_2. The overbank flow discharge Q_2 is a function of the water level, flood peak discharge and diversion angle. The channel slope of the neck position depends on the bend planform and neck width, the smaller the neck width, the larger the slope. The overbank flow in the maximum gradient direction rapidly forms a new channel in a relatively short time. During the latter half of the second stage, B_c and H may keep the same rate of increase, and the increase of B_c quickly limits the channel deepening. Meanwhile, the median grain size of the sediment becomes larger, which further inhibits the scouring, and eventually the new channel bed with the strong erosion achieves a relatively stable state.

3.3.3.2 Bank failure type

Bank failure is a natural phenomenon for nearly all alluvial rivers, especially for the outer banks of meandering rivers. It has been a long-term concern for the middle Yangtze River, the lower Mississippi River, and the Rhine River. Bank failure is affected by flow, soil characteristics, and vegetation and channel morphology. This section analyzes the bank failure type neck cutoffs using a widely cited bank stability analysis method developed by Osman and Thorne (1988).

A cross section of the river bank in a bend neck is simplified as shown in Figure 3.18. Some assumptions are listed as follows: (1) the riverbank soil is homogeneous clay soil and the bank failure plane slips through the bank toe; (2) the effects from the vegetation, soil runoff and underground water are neglected; (3) the neck bank slope is greater than 60°; and (4) the left and right banks of the meander neck are approximately symmetric and the surface tension crack is in a middle position. As shown in Figure 3.18, the lateral scour width is Δb, the lateral scouring depth is Δz, the initial channel depth is H_0, and the river depth at upstream and downstream of the neck are H_1 and H_2, respectively.

On the sliding surface CD of the left bank failure body, the resistance force can be expressed as:

$$F_{R1} = cL_{CD} + \left(W\cos\beta + F_{W1}\sin\beta + \frac{P_d b}{\sqrt{(h)^2 + b^2}}\sin\beta \right) \tan\varphi + F_{W1}\cos\beta$$

$$+ \frac{P_d b}{\sqrt{(h)^2 + b^2}}\cos\beta$$ (3.17)

where L_{CD} is the sliding surface length, W is the failure body gravity, β is the angle of sliding face, F_{W1} is water pressure of upstream and $F_{W1} = \rho g (\alpha V^2/2g + H_1^2)/2$, ρ

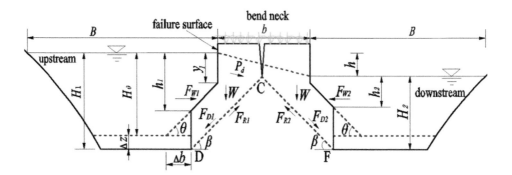

Figure 3.18 Schematic diagram of bank failure type of nect cutoff.

is the water density, g is the gravity acceleration, α is the uniformity coefficient of the velocity distribution, V is the nearbank velocity, P_d is the porous water pressure, and h is the difference of the water level on both sides of the bank.

The corresponding sliding force of the left failure body is expressed as:

$$F_{D1} = W \sin \beta + \frac{1}{2}\rho g \left(h_1^2 - y_1^2\right) \cot \theta \tag{3.18}$$

Similarly, the sliding resistance force of the right bank failure body can be expressed as:

$$F_{R2} = cL_{CF} + \left(W \cos \beta + F_{W2} \sin \beta - \frac{P_d b}{\sqrt{(h)^2 + b^2}} \sin \beta \right) \tan \phi + F_{W2}\cos \beta \tag{3.19}$$

The corresponding sliding force of the right failure body is expressed as:

$$F_{D2} = W \sin \beta + \frac{1}{2}\rho g \left(h_1^2 - y_2^2\right) \cot \theta + \frac{P_d b}{\sqrt{(h)^2 + b^2}} \cos \beta \tag{3.20}$$

Considering the difference of the sliding resistance force and slipping force, the stability of the unilateral river bank, i.e. $F_{s1} = F_{R1} - F_{D1}$ and $F_{s2} = F_{R2} - F_{D2}$, were identified as the theoretical basis to determine if the left and right bank body is stable. Eq. (3.17)–(3.20) are substituted with F_{s1} and F_{s2}. It is obvious that the left bank is more stable and the stability increases as the soil cohesion, internal friction angle and porous water pressure increase and the smaller water pressure, bank slope and sliding angle decrease. The nearbank flow scours the bank toe, resulting in suspension of the bank failure body, increase of the slope angle, and increase of the instability. The stability of the bank failure body on the right side is the same as that on the left bank except that the porous water pressure has a negative impact to the stability. Without considering the morphology of adjacent meanders, the upstream bank of the meander neck is more vulnerable to bank failure than the downstream bank.

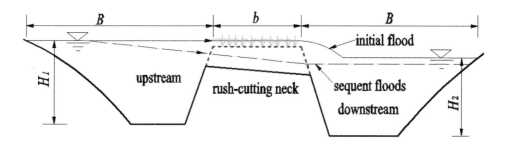

Figure 3.19 Schematic diagram of bank failure type of neck cutoff.

If $F_{s1} > 0$ and $F_{s2} > 0$, the left and right banks of the meander neck are in a steady state, and neck cutoff does not quickly occur. If $F_{s1} < 0$ and $F_{s2} < 0$, both left and right banks are in an unstable state and neck cutoff will occur sooner or later. If $F_{s1} = 0$ and $F_{s2} = 0$, both left and right banks are in a critical state and neck cutoff may occur at any time. If $F_{s1} < 0$ or $F_{s2} < 0$, a unilateral bank failure will occur at the neck. The water flow will flush the other bank and the infiltration pressures may lead to a piping effect and soil flowing, which will cause a neck cutoff to occur in a short time.

3.3.3.3   Erosion channel type

Erosion channel type neck cutoff is the result of successive overbank flow scouring during a flood. The initial flood forms the channel and subsequent floods continue to scour the previous channel until both ends of the neck intercept and the flow water interconnects (Fig. 3.19). The conditions are a wide neck and that the initial flood discharge is not strong enough to scour out a new channel. Moreover, the recession of the flood may results in stoppage of the erosion. If the vegetation coverage has strong resistance to flow and weakens the scouring ability of the flow, the erosion channel pattern will occur under subsequent overbank flow scouring.

The first step is the formation of a new channel by overbank flow scouring. The second step is that a subsequent flood comes and continuously scours the previous channel. Only when the flood level is higher than the bottom elevation of the previous channel can the overbank flow form. The formation of the previous channel reduces the difficulty of the neck cutoff formation. The scouring type is mainly headcut erosion because the neck width is quite large and the downstream water lever is below the entrance level. According to the motion equation of flow and the sediment continuity equation, the change of the channel profile of the neck channel erosion can be depicted by a classic diffusion equation.

The meandering river reaches a critical state during the occurrence of a neck cutoff, which plays a regulatory role in limiting its planform complexity and sinuosity. The traditional viewpoint considers that neck cutoffs occur during extreme floods. The overbank flow erosion by the floods is not the entire mechanism of the neck cutoff. In fact, the occurrence of a neck cutoff depends on the meander planform, riverbed morphology, extreme flood and its diversion angle, soil characteristics, and vegetation cover. The processes and mechanism of a neck cutoff interweaves a variety of factors with complexity and uncertainty.

3.4 EVOLUTION OF OXBOW LAKES

Oxbow lake, as an enclosed or semi-enclosed shallow lake, is a special kind of riparian wetland habitat, the aquatic environment and organisms of which was paid more attention to recently by researchers (Jones et al., 2008; Pan et al., 2011; Stella et al., 2011). The formation of an oxbow lake, of which the inlet and outlet withstands long-term sediment deposition, has been used to inverse the meandering river evolution, cutoff time, flood frequency, and channel sediment budget in a historical period using the sedimentology and chronology methods (Babka et al., 2011; Knight et al., 2009; Rasmussen and Mossa, 2011; Rowland et al., 2005; Thornbush and Desloges, 2011). To date, the previous studies have mainly focused on the geometric morphology, aquatic ecology and sedimentology. There is only a small amount of qualitative descriptions but theoretical analyses are still lacking on the formation and long-term evolution of oxbow lakes. This section considers a neck cutoff to be the starting point of the oxbow lake formation based on the definition of oxbow lake morphology.

An oxbow lake forms after a meander cutoff. After experiencing outlet and inlet depositions for several years, the original channel becomes a shallow and closed lake. Existence of oxbow lakes is a common landscape along river valleys, such as the middle reaches of the Yangtze River, the lower reaches of the Mississippi River, the tributaries of Yellow River source (e.g. Bai, Hei, Zequ, and Lanmucuo rivers), the rivers in Siberia alluvial plain, the Amazon River tributaries, etc. The morphology and quantity of oxbow lakes reflect the migration rate and cutoff frequency. In addition, the sedimentary structure of an oxbow lake records the deposition process and age, the flood frequency, and paleoenvironmental reconstruction.

3.4.1 Formation of oxbow lakes

3.4.1.1 Morphological parameters

This section only discusses the oxbow lake formed by neck cutoff. Oxbow lake is the inevitable production of meander neck cutoff. The formation of a meander neck cutoff consists of three stages: 1) sediment deposits at the inlet of the old channel; 2) the flood carries sediment to fill into the middle of the old channel; and 3) ultimately the inlet and outlet are blocked and an enclosed lake forms.

Figure 3.20 defines the channel morphology when a neck cutoff is occurring and an oxbow lake is forming. Assuming that the cross section is rectangular and the channel width is constant. Q is the discharge in the upstream main channel, Q_1 is the discharge in the original channel, Q_2 is the discharge in the new channel, δ is the diversion angle of the new and old channels, b is the neck width, and B is the channel width. After the neck cutoff forms, the sand plug gradually forms at the inlet section and the outlet section due to sediment deposition. L_u is the length of the sand plug at the inlet, L_d is the length of the sand plug at the outlet, Z_1 and Z_2 are the river bottom elevations at the inlet and outlet, respectively, and h_1 and h_2 are water depths at the inlet and outlet, respectively.

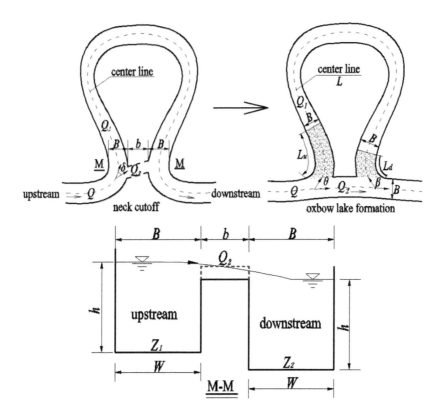

Figure 3.20 Parameters of oxbow lake.

3.4.1.2 Deposition at inlet section

If the meander development reaches a high sinuosity, the neck width contracts to quite narrow. Once the flood with high water stage scours the neck, or the two side banks of the neck collapse, or multiple overbank flow events erode the neck, the neck cutoff will finally complete. Based on his research of the Sacramento River and its oxbow lakes, Constantine et al. (2010) concluded that the diversion angle δ plays an important role on the formation rate and deposition rate of the inlet. The smaller δ is, the faster the deposition rate is. Google Earth satellite images and field observations of a large number of oxbow lakes have shown that the length and deposition rate at the inlet sections are larger than that at the outlet sections. The typical blockage process of oxbow lakes is that the inlet section is blocked first and the outlet section is blocked next. Therefore, the sediment deposition process at the inlet section is the key for analyzing the formation mechanism of oxbow lakes.

As the new channel scours and the inlet deposition of the old channel section progresses, Q_1 gradually reduces. The flow continuity equation is as follows:

$$Q_1 = U_1 h_1 B \tag{3.21}$$

For the boundary condition of the inlet section, the upstream water flow and sediment are simplified as a steady process. After the neck cutoff occurs and a new channel develops, Q_2 increases and Q_1 decreases. In order to reflect the time-dependent characteristics (decreasing with time) of Q_1, Q_1 is simplified as a linear function of attenuation as follows,

$$Q_1 = at + c \qquad (3.22)$$

where a and c are coefficients. If $t = 0$, $Q_1 = Q$ indicates that the neck is in the critical condition of a neck cutoff. If $t = t_c$, $Q_1 = 0$ indicates that an oxbow lake is formed, therefore, $a = -Q/t_c$ and $c = Q$. The derivation of Eq. (3.22) is as follows:

$$\frac{\partial Q_1}{\partial t} = -\frac{Q}{t_c} \qquad (3.23)$$

The bed deformation equation is:

$$\frac{\partial Z_1}{\partial t} = \frac{1}{\rho'} \alpha \omega \left(S - \overline{S_*} \right) \qquad (3.24)$$

The suspended load carrying capacity formula is:

$$\overline{S_*} = k \left(\frac{U_1^3}{g R_1 \omega} \right)^m \qquad (3.25)$$

There, the sediment transport continuity equation is as follows:

$$\frac{\partial (Q_1 S)}{\partial x} + \frac{\partial (A_1 S)}{\partial t} = -\alpha \omega B \left(S - \overline{S_*} \right) \qquad (3.26)$$

where x is the distance, t is time, A_1 is the sectional area, $A = Bh_1$, Z_1 is the average bed elevation, S and $\overline{S_*}$ are the average sediment concentration and sediment carrying capacity, ρ' is the sediment dry density, g is the acceleration of gravity, ω is the settling velocity, α is the recovery saturation coefficient, $\alpha = 0.25$ in deposition, $R_1 = Bh_1/(2h_1 + B)$ is hydraulic radius, and k and m are empirical coefficients.

Assuming that the suspended load concentration does not change with time, $\partial (A_1 S)/\partial t = 0$ and $dS_*/dx = 0$ will cause large errors. Eq. (3.26) belongs to the first-order linear ordinary differential equation, assuming that the settling velocity does not vary with distance x. A particular solution is obtained under the boundary condition $x = 0$:

$$S = \overline{S_*} + \left(S_1 - \overline{S_*} \right) e^{-\frac{\alpha \omega L_u}{U_1 h_1}} \qquad (3.27)$$

where S_1 and $\overline{S_*}$ are sediment concentration and sediment carrying capacity at the inlet section L_u, respectively, and L_u is the length of the sediment plug at the inlet section. Eq. (3.27) shows that the section sediment concentration consists of two parts: L_u and $(S_1 - \overline{S_*}) e^{-\frac{\alpha \omega L_u}{U_1 h_1}}$. The first term on the right side in Eq. (3.27) is the main component,

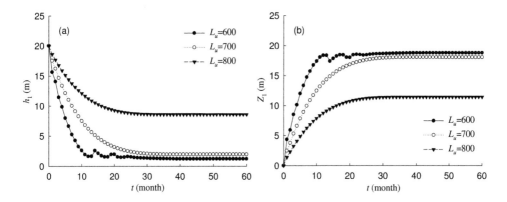

Figure 3.21 Change of water depth and bed elevation at different deposition lengths ($S_1 = 1.0$).

and the second term depends on the value of $(S_1 - \overline{S_*})$. Insert Eq. (3.27) into Eq. (3.24), and the change of the river bed elevation can be calculated as:

$$\frac{\partial Z_1}{\partial t} = \frac{\alpha \omega}{\rho'}\left(S_1 - \overline{S_*}\right)e^{-\frac{\alpha \omega L_u}{U_1 h_1}} \tag{3.28}$$

Eq. (3.22), (3.24), (3.26), and (3.28) are adopted to non-coupled solve the change of the bed elevation within L_u. Using the Shatanzi neck cutoff and oxbow lake in the middle reach of the Yangtze River as an example, the values of the variables are: $Q = 12500\,\mathrm{m^3/s^{-1}}$, $t_c = 5\,\mathrm{yr}$, $B = 1000\,\mathrm{m}$, $\rho' = 1400\,\mathrm{kg/m^3}$, $\omega = 0.284\,\mathrm{m/s^{-1}}$, $\alpha = 0.25$, $m = 0.92$, and $k = 0.03$. The initial condition is $t = 0$, $U_1 = 0.625\,\mathrm{m/s^{-1}}$, $h_1 = 20\,\mathrm{m}$, and $Z_1 = 0$. If $U_1 \approx 0$, the inlet section of the oxbow lake is silted up and the computation ends. Z_1 is the silting thickness. The time step is 1 month. If the initial and boundary conditions and the empirical coefficients have been determined, S_1 and L_u are two key factors which influence the deposition rate of the inlet section.

If $S_1 = 1.0\,\mathrm{kg\ m^{-3}}$, Figure 3.21 shows the changes of the water depth and bed elevation under the conditions of $L_u = 600\,\mathrm{m}$, $L_u = 700\,\mathrm{m}$, and $L_u = 800\,\mathrm{m}$. The figure shows the process of the decrease of the water depth and the increase of the bed elevation with a steady suspended load concentration. In Table 3.5, when $L_u = 800\,\mathrm{m}$, the water depth and bed elevation are different from the results when $L_u = 600\,\mathrm{m}$ or $L_u = 700\,\mathrm{m}$. There are two possible reasons: 1) this simplified model cannot reflect the real physical process; or 2) based on the preset parameters, the channel geometry parameters and initial conditions have different sensitivity to the sand plug with different lengths, i.e. they are more sensitive to the condition when $L_u = 800\,\mathrm{m}$. The deposition equilibrium time for $L_u = 800\,\mathrm{m}$ was 7 months shorter than that for $L_u = 700\,\mathrm{m}$. This is because the longer length of the sand bar directly leads to a reduction of the final deposition thickness and less time to reach the balance of deposition.

Figure 3.21 above shows that the deposition rate is higher during the initial 2–3 years, and as the deposition height increases, the water depth decreases. It accords with field observations that oxbow lakes deposit quickly during early years and slow

Table 3.5 Simulation results of different deposition length.

Length of sand plug (m)	Initial water depth h_1 (m)	Steady water depth h_2 (m)	Bed elevation (m)	Deposition time (month)
600	20	1.60	18.78	24
700	20	1.95	18.04	42
800	20	8.18	11.42	35

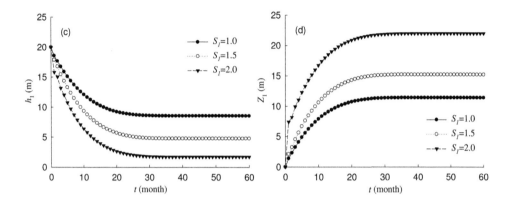

Figure 3.22 Change of water depth and bed elevation under different suspended load concentrations ($L_u = 800$).

down later. For example, the initial deposition rate of the Shatanzi oxbow lake at the inlet section was quite fast, which was 2–3 m/yr.

If $L_u = 800$ m, Figure 3.22 shows the changes of the water depth and bed elevation under the conditions of $S_1 = 1.0$, $S_1 = 1.5$, and $S_1 = 2.0\,\mathrm{kg/m^3}$. The figure shows the process of the decrease of the water depth and the increase of the bed elevation under a constant L_u. The stability time is 36 months for $S_1 = 2.0\,\mathrm{kg/m^3}$, which is longer than that for $S_1 = 1.0\,\mathrm{kg/m^3}$ as shown in Table 3.6. There are two possible reasons: 1) it is easier to form a sand plug under the condition of a higher concentration; and 2) for $S_1 = 2.0\,\mathrm{kg/m^3}$, when the water depth decreases below 2 m, the deposition rate becomes so low that it takes longer to reach equilibrium.

3.4.1.3 Deposition at outlet section

As a neck cutoff is occurring, the mainstream flows into the new channel but sediment deposits at the inlet section of the original channel, which becomes a low velocity zone so the sediment transport capacity decreases and suspended load deposits along the course. Therefore, the deposition rate of the sediment becomes quite low, which explains why the inlet section is blocked first and the outlet section is blocked later. At the outlet section, the flow with low concentration of sediment from the original channel flows into the downstream channel, where the flow regime is different from that at the inlet section. Therefore, the equations for the inlet section in the previous section cannot be used for the outlet section.

Table 3.6 Simulation results for different suspended load concentrations.

Suspended load concentration (kg/m³)	Initial water depth h₁ (m)	Steady water depth h₂ (m)	Bed elevation (m)	Deposition time (month)
1.0	20	8.58	11.42	35
1.5	20	4.80	15.20	32
2.0	20	1.72	22.00	36

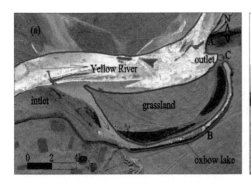

Figure 3.23 Outlet conditions of an oxbow lake on Yellow River: (a) Satellite image (33°24′N, 102°23′E); (b) Field photo in July 2011.

Field surveys were conducted to observe the flow pattern at the outlet of an oxbow lake at the Yellow River source in Tangke Town. Figure 3.23(b) shows that the flow velocity is faster and the suspended load concentration is higher in the new main channel. In the confluence between the original channel and the new channel, there is a clockwise circulation, which carries high concentration of sediment from the new channel backwards into the original channel with low concentration of sediment.

Figure 3.23(a) is a remote sensing image taken on December 18, 2010, showing the channel of the Yellow River frozen and the water drying up in the oxbow lake. Figure 3.23(b) shows the oxbow lake filling up with water during the flood season in July 2011. Point A is the location where field photos were taken, and B and C were the locations where water samples were taken. The suspended sediment concentration was 50 kg/m³ at point B, and 254 kg/m³ at point C. Meanwhile, it was observed that the mainstream turbidity was flowing into a relatively low velocity region near the outlet section and caused a clockwise circulation.

The sediment deposition at the outlet section of oxbow lakes is suspended sediment deposition during sediment-laden floods. If the discharge at the inlet is lower, the flow velocity in the lake is lower, and the deposition rate becomes larger. It is obvious that the premise of a lower discharge at the inlet section is higher bed elevation at the inlet. For instance, during the initial forming stage of the Shatanzi oxbow lake, the deposition rate at the inlet section was quite fast, up to 2–3 m/yr, while the deposition rate at other parts was 15–40 cm/yr. During the middle stage, the deposition rate was 8–25 cm/ yr at the inlet and 2.7–6.8 cm/yr at other parts.

3.4.2 Long-term evolution of oxbow lakes

Oxbow lake is an historical evolution trace of meandering rivers. The formation of oxbow lakes consists of a blockage of the inlet section first, then a blockage of the outlet section, and finally the formation of a closed lake. Once an oxbow lake is formed, it makes no movements. Only when the flow in the new channel overflows into the inlet and outlet sections that the water and sediment enters the lake again. Riparian vegetation growth, sediment deposition, evaporation and infiltration directly cause the shrinkage of an oxbow lake. Hence, the long-term evolution of an oxbow lake is mainly characterized by sediment deposition and water reduction.

To study the diversity of oxbow lakes, Weihaupt (1981) analyzed 817 oxbow lakes along the Yukon River and classified them into simple, compound and complex types based on their morphological complexity. Moreover, oxbow lakes can also be classified into open, normal, and closed types based on the closure condition. The former classification system only focuses on the geometric shape of oxbow lakes; however, the morphological complexity is merely a portrayal of the original river before the cutoff, which does not reflect the evolution of oxbow lakes. The latter classification system presents the general features of oxbow lakes in different stages, but does not reveal the specific feature of a single oxbow lake. Indeed, the different evolution stages of an oxbow lake is corresponding to the different morphology.

Oxbow lakes can also be classified based on their shapes: Ω-shape, U-shape, and Crescent-shape. Ω-shape refers to an oxbow lake that has highly curved boundaries with two ends shrinking closely. U-shape represents a shape that two ends of the oxbow lake are far apart, such as the horseshoe-shape or semi-circular shape. Crescent-shape represents a shape that the boundary line is a certain arc. The three shapes are shown in Fig. 3.24.

In order to test whether the three types are representative, the oxbow lakes along ten meandering rivers were studied using Google Earth satellite images. These ten rivers were the Bai, Hei, Irtysh, Mississippi, Red, Napo, Tigre, Purus, Jurua, and Ucayali rivers. The parameters of these rivers and the oxbow lakes are summarized in Table 3.6.

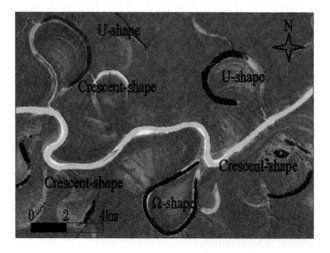

Figure 3.24 Three typical forms of oxbow lakes (from Google Earth, Jurua River, 06°35′ S, 69°15′W).

These ten meandering rivers are located in regions with different geography, climate, vegetation and human activity, representing a broad range of field conditions. The study reaches extended from the estuary upstream until the channels were not easily identified. The selection criterion was that the inlet and outlet sections were completely blocked and the oxbow lakes were filled with water. A total of 1329 oxbow lakes were identified with 349 (26%) Ω-shaped lakes, 452 (34%) U-shaped lakes, and 528 (40%) Crescent-shaped lakes, as shown in Table 3.7.

Based on the comparison of a large number of oxbow lakes included in Table 3.7, it was concluded that Ω-shape, U-shape and Crescent-shape represent three evolution stages of the entire life period of oxbow lakes. Ω-shape is the initial form of an oxbow lake, when the oxbow lake maintains the original meander shape. Along with sediment deposition, evaporation, infiltration, and vegetation encroachment, the oxbow lake gradually shrinks to U-shape, as shown in Figure 3.25. U-shape of the oxbow lake is in the middle evolution stage, which is the transition from the mature stage to perish. In this stage, the U-shape has the most abundant morphological diversity. As riparian vegetation continues to encroach the oxbow lake through sediment deposition, the oxbow lake of U-shape continues to shrink to some localized channels but still keeps a certain arc, which turns into a crescent type (Fig. 3.25).

The Crescent-shaped oxbow lake steps into the declining process due to vegetation reproduction and eventually vanishes into the land part of the alluvial plain. Ω-shape, U-shape, and Crescent-shape are a whole life cycle of an oxbow lake evolution. The premise condition of a sequential development of the three types is that a neck cutoff

Table 3.7 Classification and statistics of oxbow lakes on ten meandering rivers.

River	Location	Watershed features	Ω	U	Crescent	Number
			Type and number of oxbow lakes			
Bai	33°19'N, 102°28' E– 32°27'N, 102°22' E	Wet, grassland, weak human activity	4	29	17	50
Hei	33°58'N, 102°10' E– 32°02'N, 103°00' E	Wet, grassland, weak human activity	36	49	64	149
Irtysh	60°58'N, 69°10' E– 55°14'N, 73°07' E	Cold, floodplain, weak human activity	57	66	54	177
Mississippi	30°40'N, 91°19'W– 35°13'N, 90°08'W	Wet, floodplain, intense human activity	22	16	6	44
Red	31°13'N, 91°57'W– 33°55'N, 95°32'W	Wet, floodplain, intense human activity	23	60	26	109
Napo	02°19'S, 74°08'W– 01°35'S, 75°59'W	Tropical forest, weak human activity	54	35	38	127
Tigre	04°25'S, 74°07'W– 02°10'S, 76°01'W	Tropical forest, weak human activity	49	38	27	114
Purus	04°44'S, 62°25'W– 09°04'S, 68°15'W	Tropical forest, weak human activity	35	45	81	161
Jurua	02°41'S, 65°44'W– 07°58'S, 72°46'W	Tropical forest, weak human activity	61	100	181	342
Ucayali	04°41'S, 73°33'W– 10°00'S, 74°00'W	Tropical forest, weak human activity	8	14	34	56

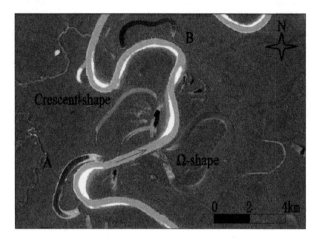

Figure 3.25 Bend migration changes the existence of oxbow lakes (from Google Earth satellite images, Purus River, 06° 47′ S, 64° 36′ W).

occurs as a Ω-shaped meander and that the new channel is away from the oxbow lake. However, during the formation of an oxbow lake, it is inevitable that the oxbow lake is affected by the evolution of the new channel, especially by the valley constraining the meander belt. Therefore, some oxbow lakes do not follow the natural order to evolve from Ω-shape, U-shape to Crescent-shape. If the migration direction of the new channel is opposite from the original channel, the oxbow lake forms and gradually moves away from the new channel, and the evolution process follows the order of Ω-shape, U-shape and Crescent-shape. If the migration direction of the new channel is towards the original channel, the formed oxbow lake has to face the concave bank of the new channel. With the new channel scouring, the oxbow lake will be encroached by the new channel as shown A and B in Figure 3.25.

In brief, there are two types of natural evolution processes of oxbow lakes. The first type is that the oxbow lake experiences Ω-shape, U-shape, and Crescent-shape, which finally becomes a part of the alluvial plain due to sediment deposition, evaporation, infiltration, and vegetation encroachment. The second type is that the oxbow lake is gradually eroded by the new channel and becomes a part of the alluvial river and plain, and completely loses its original trace.

REFERENCES

Abernathy B., Rurtherford I.D. 2001. The distribution and strength of riparian tree roots in relation to riverbank reinforcement. Hydrological Processes, 15:63–79.

Asahi K., Shimizu Y., Nelson J., Parker G. 2013. Numerical simulation of river meandering with self-evolving banks. Journal of Geophysical Research: Earth Surface, 118:2208–2229.

Babka B., Futo I., Szabo S. 2011. Clustering oxbow lakes in the Upper-Tisza Region on the basis of stable isotope measurements. Journal of Hydrology, 410(1–2):105–113.

Blondeaux P., Seminara G. 1985. A unified bar-bend theory of river meanders. Journal of Fluid Mechanism, 157:449–470.

Braudrick C.A., Dietrich W.E., Leverich G.T., Sklar L.S. 2009. Experimental evidence for the conditions necessary to sustain meandering in coarse-bedded rivers. PNAS, 106(40):16936–16941.

Brooks G.R., Medioli B.E. 2003. Deposits and cutoff ages of horseshoes and Marion oxbow lakes, Red River, Manitoba. Geographie Physique and Quaternarie, 57(2–3):151–158.

Burckhardt J.C., Todd B.L. 1998. Riparian forest effect on lateral stream channel migration in the glacial till plains. Journal of the American water resource association, 34:179–184.

Camporeale C., Perona P., Porporato A., Ridolfi L. 2005. On the long-term behavior of meandering rivers. Water Resources Research, 41, W12403, doi: 10, 1029/2005WR004109.

Camporeale C., Perucca E., Ridolfi L. 2008. Significance of cutoff in meandering river dynamics. Journal of Geophysical Research, 113, F01001, doi:10.1029/2006JF000694.

Constantine J.A., Dunne T. 2008. Meaner cutoff and the controls on the production of oxbow lakes. Geology, 36(1), 23–26.

Constantine J.A., Dunne T., Piegay H., Kondolf G.M. 2010. Controls on the alleviation of oxbow lakes by bed-material load along the Sacramento River, California. Sedimentology, 57:389–407.

Constantine J.A., Mclean S.R., Dunne T. 2010. A mechanism of chute cutoff along large meandering rivers with uniform floodplain topography. GSA Bulletin, 122(5/6):855–869.

Da Silva A.M.F. 2006. On why and how do rivers meander. Journal of Hydraulic Research, 44(5):579–590.

Dietrich W.E., Smith J.D. 1983. Influence of the point bar on flow through curved channels. Water Resources Research, 19:1173–1192.

Dijk W.M., Lageweg W.I., Kleinhans M.G. 2012. Experimental meandering river with chute cutoffs. Journal of Geophysical Research, 117, F03023, doi: 10.1029/2011JF002314.

Duan J.G., Julien P.Y. 2010. Numerical simulation of meandering evolution. Journal of Hydrology, 391:34–46.

Dulal K. P., Kobayashi K., Shimizu Y., Parker G. 2010. Numerical computation of free meandering channels with the application of slump blocks on the outer bends. Journal of Hydro-environment Research, 3:239–246.

Eke E., Parker G., Shimizu Y. 2014. Numerical modeling of erosional and depositional bank processes in migrating river bends with self-formed width: morphodynamics of bar push and bank pull. Journal of Geophysical Research: Earth Surface, DOI:10.1002/2013JF003020.

Erskine W., Mcfaden C., Bishop P. 1992. Alluvial cutoffs as indicators of former channel conditions. Earth Surface Processes and Landforms, 17(1): 23–37.

Erskine W., Melville M., Page K.J., et al. 1982. Cutoff and oxbow lake. Australian Geographer, 1982, 15:174–180.

Fuller I.C., Large A.R.G., Milan D.J. 2003. Quantifying channel development and sediment transfer following chute cutoff in a wandering gravel-bed river. Geomorphology, 54: 307–323.

Gay G.R., Gay H.H., Gay W.H., Martinson H.A., Meade R.H. 1998. Evolution of cutoffs across meander necks in Power River, Montana, USA. Earth Surface Processes and Landforms, 23(7):651–662.

Greg E., Lance D.Y. 2002. The effects of riparian vegetation on bank stability. Environmental and Engineering Geoscience, 8(4):247–260.

Ghinassi M. 2011. Chute channels in the Holocene high-sinuosity river deposits of the Firenze plain, Tuscany, Italy. Sedimentology, 58:616–642.

Grenfell M., Aalto R., Nicholas A. 2012. Chute channel dynamics in large, sand-bed meandering rivers. Earth Surface Processes and Landforms, 37(3):315–331.

Guneralp I., Rhoads B.L. 2009. Empirical analysis of the planform curvature-migration relation of meandering rivers. Water Resources Research, 45, W09424. DOI:10.1029/2008/WR007533.

Hager W.H. 2003. Fargue, founder of experimental river engineering. Journal of Hydraulic Research, 41(3):227–233.

Handy R.L. 1972. Alluvial cutoff dating from subsequent growth of a meander. Geological Society of America Bulletin, 83:475–480.

Hickin E.J., Nanson G.C. 1984. Lateral migration rates of river bends. Journal of Hydraulic Engineering, 110:1557–1567.

Hooke J.M. 2004. Cutoffs galore!:Occurrence and causes of multiple cutoffs on a meandering river. Geomorphology, 61(3/4):225–238.

Hooke J.M. 2007. Complexity, self-organization and variation in behavior in meandering rivers. Geomorphology, 91:236–258.

Hooke J.M. 1995. River channel adjustment to meander cutoffs on the River Bollin and River Dane, northwest England. Geomorphology, 14(3):235–253.

Howard A.D. 1984. Sufficient condition for meandering:a simulation approach. Water Resources Research, 20:1659–1667.

Howard A.D. 1992. Modeling channel migration and floodplain sedimentation in meandering streams. Lowland Floodplain Rivers:Geomorphological Perspectives. Edited by P. A. Carling and G. E. Petts, John Wiley & Sons Ltd, 1–41.

Howard A.D. 2009. How to make a meandering river. PNAS, 106(41):17245–17246.

Hubble T.C.T., Docker B.B., Rutherfurd I.D. 2010. The role of riparian trees in maintaining riverbank stability:A review of Australian experience and practice.Ecological Engineering, 38:292–304.

Ikeda S., Parker G., Sawai K. 1981. Bend theory of river meanders. Part 1. Linear development. Journal of Fluid Mechanism, 112:363–377.

Jones J.R., Obrecht D.V., Perkins B.D., Knowlton M.F., Thorpe A.P., Watanabe S., Bacon R.R. 2008. Nutrients, section, and transparency of Missouri reservoirs and oxbow lakes:An analysis of regional limnology. Lake and Reservoir Management, 2008, 24(2): 155–180.

Lancaster S.T., Bras R.L. 2002. A simple model of river meandering and its comparison to natural channels. Hydrological Processes, 16:1–26.

Langbein W.B., Leopold L.B. 1966. River meanders-theory of minimum variance. Geological Survey Professional Paper, 422-H.

Leopold L.B., Langbein W.B. 1960. River Meanders, Geological Society of America Bulletin, 71(6):769–793.

Lewis G.W., Lewin J. 1983. Alluvial cutoffs in Wales and the Borderlands. Spec. Publs. Int. Ass. Sediment, 6:145–154.

Lonsdale P., Hollister C.D. 1979. Cut-offs at an abyssal meaner south of Iceland. Geology, 7:597–601.

Kinoshita R. 1961. Investigation of channel deformation in Ishikari River. Report to the Bureau of Resources, 1–174.

Knight S.S., Lizotte R.E., Moore M.T., Smith F., Shields F.D. 2009. Mississippi oxbow lake sediment quality during an artificial flood. Bull. Environ. Contam. Toxicol., 82(4): 496–500.

Kort J., Collins M., Ditsch D. 1998. A review of soil erosion potential associated with biomass crops. Biomass and Bioenergy, 14:351–359.

Micheli E.R., Kirchner J.W., Larsen E.W. 2002. Quantifying the effect of riparian forest versus agricultural vegetation on river meander migration rates, central Sacramento river, California, USA. River research and applications, 20:537–548.

Micheli E.R., Larsen E.W. 2011. River channel cutoff dynamics, Sacramento River, California, USA. River Research and Applications, 27(3):328–344.

Motta D., Abad J.D., Langendoen E.J., Garcia M.H. 2012. A simplified 2D model for meander migration with physically-based bank evolution. Geomorphology, 163/164:10–25.

Nicoll T.J., Hickin E.J. 2010. Planform geometry and channel migration of confined meandering rivers on the Canadian prairies. Geomorphology, 116:37–47.

Osman A.M., Thorne C.R. 1988. Riverbank stability analysis I: Theory. Journal of Hydraulic Engineering, 114(2):134–150.

Pan Baozhu, Wang Zhaoyin, Xu Mengzhen. 2011. Macroinvertebrates in abandoned channels assemblage characteristics and their indications for channel management. River Research and Applications, 28(8):1149–1160.

Pan Qingshen, Shi Shaoquan, Duan Wenzhong. 1978. A study of the channel development after the completion of artificial cutoffs in the middle Yangtze River. Scientia Sinica, 21(6): 37–49.

Parker G., Shimizu Y., Wilkerson G.V., Eke E.C., Abad J.D., Lauer J.W., Paola C., Dietrich W.E., Voller V. R. 2011. A new framework for modeling the migration of meandering rivers. Earth Surface Processes and Landforms, 36(1):70–86.

Pittaluga M.B., Nobile G., Seminara G. 2009. A nonlinear model for river meandering. Water Resources Research, 45, W04432, doi: 10.1029/2008WR007298.

Posner A.J., Duan J.G. 2012. Simulating river meandering processes using stochastic bank erosion coefficient. Geomorphology, 163/164:26–36.

Prosser I.P., Dietrich W.E., Stevenson J. 1995. Flow resistance and sediment transport by concentrated overland flow in a grassland valley. Geomorphology, 13:508–519.

Rasmussen J., Mossa J. 2011. Oxbow lakes as indicators of river channel change: Leaf River, Mississippi, USA. Physical Geography, 32(6):497–511.

Rowland J.C., Lepper K., Dietrich W.E., Wilson C.J., Sheldon R. 2005. Tie channel sedimentation rates, oxbow formation age and channel migration rate from optically stimulated luminescence(OSL) analysis of floodplain deposits. Earth Surface Processes and Landforms, 30(9):1161–1179.

Smith L.M., Winkley B.R. 1996. The response of the Lower Mississippi River to river engineering. Engineering Geology, 45:433–455.

Stølum H.H. 1996. River meandering as a self-organization process. Science, 271(5256):1710–1713.

Stølum H.H., Friend P.F. 1997. Percolation theory applied to simulated meander belt sandbodies. Earth and Planetary Science Letters, 153:265–277.

Stella J.C., Hayden M.K., Battles J.J., Piegay H., Dufour S., Fermier A.K. 2011. The role of abandoned channels as refugia for sustaining pioneer riparian forest ecosystems. Ecosystems, 14(5):776–790.

Sun T., Meakin P., Jossang T., Schwarz K. 1996. A simulation model for meandering rivers. Water Resources Research, 32(9):2937–2954.

Thompson D.M. 2003. A geomorphic explanation for a meander cutoff following channel relocation of a coarse-bedded river. Environmental Management, 31(3): 383–400.

Thornbush M.J., Desloges J.R. 2011. Palaeo-environmental reconstruction at an oxbow lake situated at the lower Nottawasaga River, Southern Ontario, Canada. Environmental Archaeology, 16(1): 1–15.

Wang Z.Y., Huang J.C., Su D.H. 1997. Scour rate formula. International Journal of Sediment Research, (3): 11–20.

Weihaupt J.G. 1977. Morphometric definitions and classifications of oxbow lakes, Yukon River basin, Alaska. Water Resources Research, 13(1):195–196, doi: 10.1029/WR013i001p00195.

Wu T.H., Mckinnell W.P., Swanston D.N. 1979. Strength of tree roots and landslides on Prince of Wales Island, Alaska. Canadian Geotechnical Journal, 16(1):19–33.

Xiong Shaolong, Xu Zhaocai, Wang Wenjie. 2002. New method for meander-loop cutoffs. Journal of Hydraulic Engineering, 128(3): 354–358.

Ying Xueliang. 1965. A preliminary experiment on formation reason and fluvial processes of meandering river. Acta Geographica Sinica, 31(4):288–303. (in Chinese)

Zhao Z. 2005. Flow and sediment transport of Bai and Hei rivers in the upper Yellow River. Water Resources and Hydropower of Gansu, 27(2):153–160. (In Chinese)

Zhang B., Ai N., Huang Z.W., Yi C.B., Qin F.C. 2008. Meanders of the Jialing River in China: morphology and formation. Chinese Science Bulletin, 53(2): 267–281.

Zinger J.A., Rhoads B.L., Best J.L. 2011. Extreme sediment pulses generated by bend cutoffs along a large meandering river. Nature Geoscience, 4(10): 675–678.

Chapter 4

Wetlands and wetland shrinkage

4.1 WETLANDS IN YELLOW RIVER SOURCE

Wetlands provide habitat for diverse wildlife, act as carbon sinks that mitigate effects of climate warming, and perform important ecosystem services through their role as natural 'filters' that protect water supply and buffer nutrient loads (Barbier et al., 1997; Brinson and Malvarez, 2002). Across the world, widespread loss and degradation of terrestrial and aquatic ecosystems has intensified in the twentieth century. Recently, it is reported that the total area of wetlands in China in 2008 was around 324,097 km² (35% of which was made up by swamp or marshes, and 26% of which was made up by lakes), a 33% reduction since 1978 (Niu et al., 2012). The rapid shrinkage of wetland areas in China is mainly attributed to fill, drainage, and agriculture development associated with population growth and rapid economic development and climate warming (Zhou et al., 2009; Qiu et al., 2011). Enhanced wetland protection and restoration is an important component of aquatic and terrestrial ecosystem maintenance, flood mitigation, and water quality improvement (Cao and Fox, 2009; Niu et al., 2011).

The wetlands in the Yellow River source have diverse features due to the geodiveristy in this area (Blue et al., 2013; Nicoll et al., 2013; Gao et al., 2013). Besides the river wetland type, there are five other wetland types: alpine meadows, lakes, oxbow lakes, crescent lakes, and swamp.

(1) Alpine meadows

An alpine meadow refers to the lowland near the gentle slope in an alpine watershed. In an alpine meadow, snowmelt and rainfall produce groundwater seepage and form shallow water zones, where the water is stagnant and the flow velocity is very slow due to narrow outlets or blocked drainage. The primary vegetation type on alpine meadows is perennial herbaceous phreatophyte which typically grows in areas with high moisture contents. Star Sea in the Yellow River source is a typical alpine meadow. The Yellow River source is in an area with arid-cold and even deserted climate with a lack of precipitation, however, evaporation is relatively low in this area. Under these conditions, many alpine meadows form on low-lying lands with surface runoff and high groundwater levels. Plants on the alpine meadows include *Kobresia Schoenoides, Kobrbresiapygmaea, Kobresia Humilis, Gramineae Poaceae, Cyperaceae, Medulla Junci, Leguminosae,* and broad-leaved herb.

Alpine meadows in the Yellow River source has shrunk rapidly in the past fifty years. The main reason for the shrinkage is the drainage facilities as a result of highway

(a)

(b)

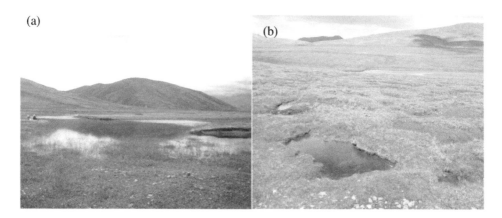

Figure 4.1 Two typical alpine meadows (a) in 2007 and (b) in 2013 (severely unwatered due to increased drainage).

constructions. Figure 4.1 shows two typical alpine meadows in the Yellow River source in 2007 and 2013. The outlet drainage conditions of many meadows have been modified and drainage rate greatly increased due to highway construction and grassland expansion. As a result, the areas that had been submerged by shallow water for a long time in these meadows shrunk significantly. On one hand, this change has benefited the grazing pastures and highway construction. On the other hand, the area of meadow wetlands has greatly reduced and caused meadow vegetation to transform to grassland vegetation.

(2) Lakes

There are many lakes in the Yellow River source, with the largest ones being the Zhaling Lake and the Eling Lake. There are thirty-one lakes with surface areas larger than 1 km², such as Xingxinghai (including four lakes), Chamucuao and Zhuorangcuo. During 2012–2013, the total area of all the lakes in the Yellow River source was estimated to be 1,456 km², of which the Eling and Zhaling lakes accounted for more than three-fourths. However, another report showed that the total area of lakes was 1,177 km² (Shang et al., 2006). The difference might have been caused by outdated data and incomplete statistics.

The surface area of the Zhaling Lake is 526 km² with an average water depth of 8.6 m. The shallowest location has a water depth of only 1 m. The storage capacity of the Zhaling Lake is approximately 4.6 billion m³. The Yellow River flows through the Zhaling Lake along a 20-km-long and 300-m-wide river valley, entering into the Eling Lake at the southwest corner. The multi-year mean annual runoff of the Yellow River at the Huangheyuan hydrological station near Maduo County is 485 million m³/yr. The surface area of the Eling Lake is 628 km² with an average water depth of 17.6 m and a maximum water depth of 30 m. The storage capacity of the Eling Lake is approximately 10.7 billion m³.

On the south of the Yellow River and the east of the Eling Lake, across the Yema ridge, four large lakes including Longrecuo (19 km²), Arong Gongmacuo (30 km²),

(a) (b)

Figure 4.2 (a) An oxbow lake created by the Yellow River near Kesheng Town; (b) An oxbow lake of the Yellow River near Tangke town with rapid siltation.

Arong Wamacuo (31 km^2), and Arong Gamacuo (23 km^2), and many small lakes form the Xingxinghai Wetland. Unlike the Zhaling and Eling lakes, these four lakes are not connected with the Yellow River. The combined area of the Zhaling Lake, the Eling Lake and Xingxinghai has increased slightly in the past 20 years according to field and remote sensing images measurements, but the total area of the lakes in the Yellow River source has had no obvious change. The area of some lakes increased due to snow-glacier melting or dam constructions, but the area of some other lakes decreased slightly due to the increase of drainage.

(3) Oxbow lakes

There are numerous oxbow lakes along the Yellow River and tributaries. These oxbow lakes are the residues of meander cutoffs from the migration of meandering rivers. Figure 4.2(a) shows a large oxbow lake near the Yellow River near Kesheng Town, which is 1 km long, 100 m wide and covered with aquatic plants. Figure 4.2(b) shows another large oxbow lake within the Yellow River First Bend near Tangke Town, which is 3 km long, 100 m wide, with shallow water depth and connected with the river.

Many rivers in the Yellow River source are meandering rivers, which have produced many oxbow lakes. For instance, the Bai River has more than 90 oxbow lakes along its course and the Hei River has more than 250 oxbow lakes. Figure 4.3(a) shows the plane shape of the Lanmucuo River with several oxbow lakes in the far distance. Figure 4.3(b) shows an older oxbow lake which was formed a long time ago and has been partially filled up by deposition and covered by vegetation.

(4) Crescent lakes

As many people know, there is a famous Crescent Lake in the Dunhuang Desert, Gansu Province. The lake is nearly 100 m long and 25 m wide with a maximum water depth of about 5 m. It has a curved shape like the crescent and is called the First Desert Spring. Several different theories have been developed to explain the formation of crescent lakes. The common point of all these theories is that the ground water level rises to be

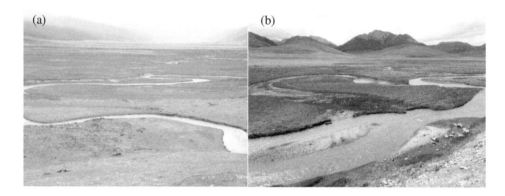

Figure 4.3 (a) Several oxbow lakes along the Lanmucuo River; and (b) An oxbow lake which is partially filled with sediment and covered by aquatic vegetation.

Figure 4.4 Formation of two crescent lakes in the Maduo Desert.

higher than the crescent bottom of sand dunes and forms crescent lakes. The crescent lakes are continuously replenished by ground water and do not dry up. In fact, the formation of crescent lakes is a distinctive hydrogeomorphic event. As long as the ground water level in the desert exceeds the elevation of the dune depressions, small lakes will form at these locations. Since sand dune depressions are typically in crescent shapes, these small lakes are called crescent lakes.

In summer of 2011, our research team identified many small crescent lakes in the Maduo Desert between the Eling Lake and the Yellow River (Wang and Han, 2015). There are more than 8,400 crescent lakes in the deserts of the Yellow River source based on investigations using remote sensing images. The dimensions of these crescent lakes are 100–300 m in length, 30–100 m in width and 1–10 m in depth. Figure 4.4 shows two crescent lakes under early stages of development. Mountain glacial and snow melting due to climate warming since 1960 have resulted in the rising of the groundwater level. The groundwater penetrates through the sand into the depressions between the dunes, forming many crescent lakes. In 2002, the construction of the Huangheyuan Dam at downstream of the Eling Lake was completed. The dam is 20 m high. Based on analysis of satellite images, the number of crescent lakes have increased from just over 6,800 to more than 8,400 since the dam was built, and the total surface area of crescent lakes also increased significantly. Crescent lake wetland is a newly-found wetland type on the Qinghai-Tibet Plateau. The formation, evolution, stabilization of sand, and ecological effects of crescent lakes wetland are very important.

(5) Swamp
Swamp in the Yellow River source mainly reside in the Ruoergai Basin. The difference between a swamp and a meadow lies in whether the surface water is flowing or not. In a meadow, the water flows slowly through shallow flooded zones. While in a swamp, the water is typically stagnant. Because of the long-term hydrostatic inundation and lower hypoxic or anoxic conditions, dead aquatic plants in swamp cannot be oxidized and they become a peat layer at the bottom of the swamp. The thickness of peat layer in the Ruoergai Swamp is in the range of 0.1–5 m.

The Ruoergai Swamp is the most important wetland in the Yellow River source. The Ruoergai (Zoige) Basin is located in the source zone of the Yellow River, at the eastern margin of the plateau. It covers a total area of 16,000 km^2, about 80% of which is in Sichuan Province and 20% within Qinghai and Gansu provinces. The Ruoergai Swamp within the Ruoergai Basin is the largest plateau peat swamp in the world. It acts as a natural reservoir that regulates the water supply to the upper Yellow River, supplying 30 % of the annual runoff at the Maqu hydrological station.

The five wetland types described above in the Yellow River source have experienced various degrees of changes over the last several decades. These changes are summarized as follows: (1) the alpine meadows and swamp have been shrinking rapidly and imposing great impacts on the plateau ecosystem; (2) the lakes have changed little; (3) the oxbow lakes are currently keeping a dynamic equilibrium with periodical generation and disappearing; (4) the crescent lakes are a new-born wetland group due to the rising of the local groundwater level, which plays an immeasurable role in stabilizing and greening deserts; and (5) the swamp are experiencing shrinkage and ecological degradation due to upstreamward incision of river network and artificial ditches draining water.

4.2   SHRINKAGE OF THE RUOERGAI SWAMP

4.2.1   The Ruoergai Basin

The Ruoergai Basin (31°51′–34°48′N, 100°46′–103°39′E) is located in the Yellow River source at elevations 3,400–3,900 m (Fig. 4.5). The swampland is a widespread development between low hills in the Ruoergai Basin. The swampland covers five counties: Hongyuan, Ruoergai, and Aba counties in Sichuan Province, and Maqu and Luqu counties in Gansu Province.

Although the plateau reached near its present elevations 13–14 million years ago, deformation within this area has continued till today as a result of the ongoing northward movement (approximately 50 mm/yr) of the Indian Plate (DeMets et al., 1994). The uplift of the plateau was accompanied by the development of a series of large

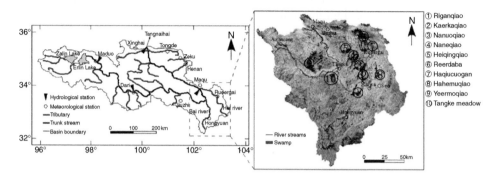

Figure 4.5 (a) Location of the Ruoergai (Zoige) Swampland in the upper Yellow River Basin, and (b) Remote sensing image of the Ruoergai Basin and ten typical swamps.

strike-slip faults and associated extensional normal faulting. The current slip rates on these faults average 1–20 mm/yr (Tapponnier et al., 2001). Many of the sedimentary basins on the plateau, such as the Ruoergai Basin, are related to these fault systems (Fu and Awata, 2007). These sedimentary basins are currently separated from each other by actively growing mountain ranges (Craddock et al., 2010, Perrineau et al. 2011). The contemporary low-relief landscape of the Ruoergai Basin was created through gradual infilling of the basin over time, accumulating over 300 m of lacustrine and fluvial silts over the past 0.9 million years (Wang et al., 1995).

The transition of the Ruoergai Basin occurred recently, changing from a gradually infilling internally-drained basin to the present external drainage. Fluvial incision of the upper Yellow River began on the northeastern margin of the Plateau since 1.8 million years ago (Li et al., 1997). The headcut erosion continued upstream at a rate of roughly 350 km M/yr (Craddock et al., 2010; Harkins et al., 2007), with a mean incision rate of 4 mm/yr at the Longyang Gorge section within the Gonghe Basin (Perrineau et al., 2011). The upper Yellow River dissected the ancient lake at Ruoergai around 38~35 kyr ago (Wang et al., 1995). These marked landscape changes have been accompanied by significant long-term climate changes. Craddock et al. (2010) concluded that the climatic conditions during the Quaternary promoted lake expansion, resulting in lake spillover. The conditions were mainly dry and cold around 30~10 kyr ago, transitioning into cool and humid conditions in the early Holocene, with precipitation decreasing slightly (promoting dry conditions) since 7.9–5.5 kyr ago (Wang et al., 1996).

The Yellow River flows through the basin in a skewed "U" shape (Li et al., 2013). This "U" shape bend is a product of the tectonic controls upon the basin formation and incisional history of the Ruoergai Basin, potentially a result of the deformation related to the Kunlun fault complex (Harkins et al., 2007). Although the Yellow River has incised through the basin fill at the northern edge of the Ruoergai (Zoige) Basin, it remains unclear whether the incision was linked to the changing base levels in the incisional landscapes of the sedimentary basins downstream (Tongde and Gonghe) or it was a product of the local tectonic deformation (Harkins et al., 2007, Perrineau et al., 2011). Recently, ^{14}C and OSL chronologies derived by Harkins et al. (2007)

and Craddock et al. (2010) indicated that the transient and rapid fluvial incision of the upper Yellow River has occurred at a rate of about 350 km M/yr. The average riverbed incision rate in the Maqu River section is about 0.5–1 m k/yr.

The Bai and Hei rivers drain parts of the Ruoergai Swamp. The Bai and Hei rivers and their tributaries meander freely within the wide piedmont valley of the Ruoergai Basin (Li et al., 2013). Numerous terraces, oxbow lakes, and abandoned channels reflect various phases of aggradation/incision, channel migration, cut-off activity, and avulsion.

Due to its location, the Maqu hydrological station on the Yellow River trunk stream provides direct measurements of the water input from the Bai and Hei rivers. The Ruoergai swamp feeds 30% of the average annual runoff of the Yellow River at the Maqu station (around 4.61×10^9 m^3/yr).

4.2.2   Shrinkage of the Ruoergai Swamp

Prior to the 1950s, the Ruoergai Swamp covered an area greater than 4,600 km^2. Its present area is just 2,200 km^2, a loss of around 52.2%, seriously degrading the functionality of this system (Yan and Wu, 2005; Li et al., 2011). Historical records and documents indicated that, prior to the 1930s, the swamp had a near-pristine landscape which experienced negligible human activities such as herding. Access to the main swamp was restricted because of the over 1 m-deep perennial sheet water. The Ruoergai Swamp has an infamous place in its history as more than 23,300 members of the Chinese Red Army died while trying to walk through/across the swamp during the rainy season in August 1935 and July 1936 (Baidu Encyclopedia, 2013). Harsh physical conditions were accompanied by food shortage and starvation. Today, much of the swamp has become grasslands, and some local areas are even prone to desertification. Photographic images taken in the early 1960s show evidence of shrinkage and degradation of the Ruoergai Swamp, with marked reduction in the total area of the surface water. Declining groundwater level was also experienced during this time period. Remote sensing images indicate even more rapid shrinkage of the swamp in the 1980s. After 2000, other than the large areas of the lakes and Swamp along the upper and middle sections of the Hei River, the rest of this area have been transformed into 'wet grassland', with no distinct surface water across the swamp even during the rainy season (Yang, 1999; Shen et al., 2003). In summary, over the past 60 years, the area of the Ruoergai Swamp has decreased by more than 52%, with serious desertification along both banks of the upper Yellow River and along the Hei and Bai rivers (Fig. 4.6).

4.2.3   Shrinkage mechanisms of the Ruoergai Swamp

In order to investigate the mechanisms for the shrinkage of the Ruoergai Swamp, detailed field investigations and measurements were completed during the summers of 2011–2013. Hydrological data were collected from the Yellow River hydrological yearbooks. The water level (stage) and average discharge records from the Tangnaihai (1959–1990) and Maqu (1959–1971) hydrological stations were analysed. Statistical analyses of the monthly temperature and precipitation were performed for Maqu County (1957–2011), Hongyuan County (1961–2011), and Ruoergai County (1967–2011). The seasons were defined as follows: spring was March–May, summer was

Figure 4.6 Shrinking swamp and desertification after drainage in the Ruoergai Swamp.

June through August, autumn was September through November, and winter was December through February. A linear trend analysis was completed to assess the inter-annual variability in the average monthly temperature and precipitation for summer and autumn.

Remote sensing analyses were completed using the Landsat Thematic Mapper (30 m resolution) and Google Earth images (65 cm panchromatic at nadir, 2.62 m multispectral at nadir) downloaded from the computer network information center, Chinese Academy of Sciences international scientific data mirror site (http://datamirror.csdb.cn).

Artificial channels excavated within the swamp from the 1960s to the 1990s were easily identified in the field as they were generally straight with relatively uniform width (typically 0.5–2.0 m width). These artificial channels are typically connected with the 'natural' streams. The characteristics of the artificial channels readily supported the identification and mapping of these channels using remotely sensed images. The particle sizes of the bed and bank materials were analyzed using a Mastersizer 2000 laser particle size analysis machine. The organic content of four peat samples was measured using the 9 TOC-V WP in the Center for Environmental Quality Test of Tsinghua University.

4.2.3.1 Climate change

Some researchers believe that the shrinkage of the Ruoergai Swamp in recent decades is a result of global climate warming, inferring that the higher temperature has increased evaporative losses (e.g. Yan and Wu, 2005). However, Ning et al. (2011) pointed out that the evaporation rate at the Maqu weather station showed a decreasing linear trend of 44.0 mm/10 yr from 1969–2008. Therefore, whether the slowly rising temperature in the Ruoergai area has positive or negative effects is still controversial.

Linear trends for the summer and autumn temperature changes in the Ruoergai area are shown from 1957–2011 in Figure 4.7. Decadal increases of 0.75°C/10 yr and 0.95°C/10 yr, respectively, are evident in Ruoergai County (Fig. 4.7a). Corresponding values for the Hongyuan and Maqu Counties are 0.21°C/10 yr and 0.27°C/10 yr, and 0.38°C/10 yr and 0.46°C/10 yr, respectively (Fig. 4.7b and 4.7c).

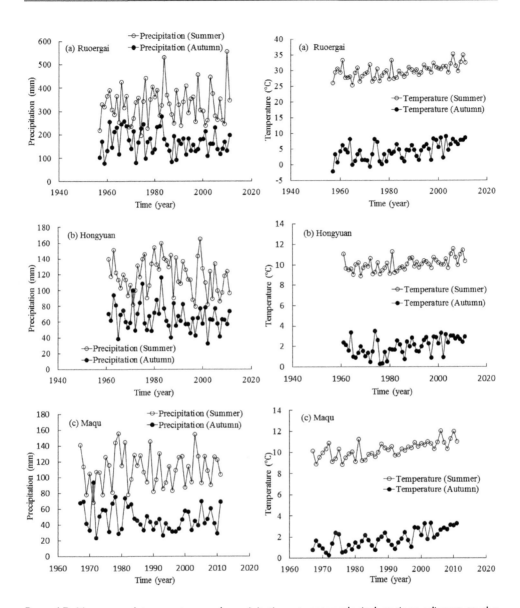

Figure 4.7 Mean annual temperature and precipitation at meteorological stations adjacent to the Ruoergai Swamp: (a) Ruoergai County (1957–2011); (b) Hongyuan County (1961–2011); (c) Maqu County (1967–2011).

The Ruoergai area has a long cold winter season. The swamp is most promi-nent in summer and autumn because of seasonal precipitation.Summer and autumn (June through November) account for 77.2%, 74.4%, and 79.4% of the annual pre-cipitation at Ruoergai, Hongyuan, and Maqu, respectively. Over the past 50 years, the summer precipitation at Ruoergai and Maqu has increased slightly (Fig. 4.7a and

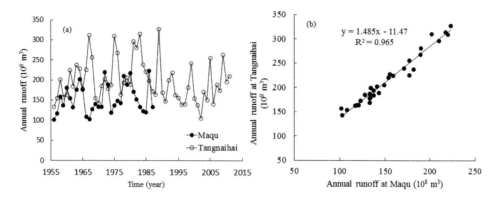

Figure 4.8   (a) Annual runoff at Tangnaihai and Maqu stations (1956–2011); and (b) Annual runoff relation between Tangnaihai and Maqu stations (1956–2011).

4.7c), but it has decreased slightly at Hongyuan (Fig. 4.7b). In all three counties, the autumn rainfall has slightly decreased (Fig. 4.7a to 4.7c). None of these changes are statistically significant, indicating that the annual precipitation at the Ruoergai Swamp remained nearly consistent from 1957–2011. Hence, the precipitation is not considered to be a contributing factor for the swamp shrinkage. Similarly, limited changes in the temperature since 1957 do not contribute to the shrinkage of the swamp (Fig. 4.7).

4.2.3.2   *Runoff reduction*

At the Tangnaihai station (1956–2011), the average annual runoff during 1991–2011 was 212.3 billion m³, which was 17.83 billion m³ less than that during 1956–1990 (21.21×10^9 m³), a 15.9% reduction (Fig. 4.8a). Since the annual runoff at the Maqu and Tangnaihai stations has a very high correlation coefficient of 0.965 (Fig. 4.8b), the annual runoff at the Maqu station was estimated based on the runoff data at the Tangnaihai station. A very high correlation ($R^2 = 0.963$) between the water stage and the annual runoff is also evident at Maqu (Fig. 4.9). Extrapolating the trend of the annual runoff at Tangnaihai to the Maqu station, a water stage drop of 0.16 m from 1991–2011 was indicated. The base levels of water stage in the Bai and Hei rivers have decreased by a similar amount.

Runoffs in the Bai and Hei rivers decreased from 1981–2002 at rates of 0.398×10^6 m³ yr^{-1} and 0.514×10^6 m³ yr^{-1}, respectively (Fig. 4.10). Over this 20 year period, the annual runoff in the upper Yellow River also decreased notably. The reduction in the runoffs was an indication of the reduction of the swamp's ability to store water and replenish the rivers. As the runoffs decreased, the water levels in the rivers also decreased, which further aggregated the shrinkage of the swamp. The lowering of the water level in the swamp reduced the groundwater level at the Ruoergai Swamp, encouraging rapid shrinkage of the swamp (as noted by local herders; Zhang and Lu, 2010).

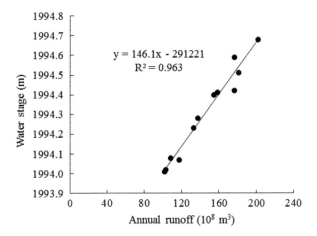

Figure 4.9 Stage-runoff relation at Maqu station (1959–1990).

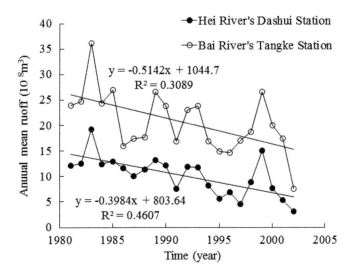

Figure 4.10 Decreasing runoff of Bai River at Tangke Station and Hei River at Dashui Station from 1981–2002.

4.2.3.3 Artificial drainage

Hundreds of kilometers of ditches were excavated in Ruoergai, Hongyuan, and Maqu counties from the 1960s to the 1990s in efforts to quickly convert swampy lands to grassland to support cattle husbandry. Official records indicate that 380 km of ditches were excavated in Ruoergai County from 1965–1973, directly draining about 800 km² of the perennial swamp (Gao, 2006). Similar figures show that 50.5 km of ditches converted 148 km² of swamp to grassland in the 1990s (Shen et al., 2003). Today, the total length of artificial ditches in Ruoergai and Hongyuan counties exceeds 1,000 km,

Figure 4.11 Artificial ditches excavated in the Ruoergai Swamp.

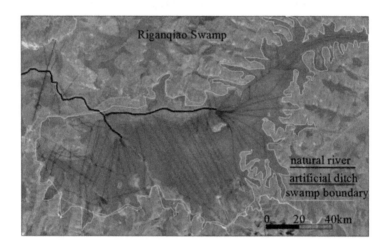

Figure 4.12 Artificial ditches (red solid lines) at the Riganqiao Swamp (Google Earth images, September 20, 2010).

draining about 2,000 km² (about 43.5% of the total area) of the swamp (Yang, 1999). Figure 4.11 shows typical artificial ditches in the Ruoergai Swamp. Figure 4.12 shows the Riganqiao Swamp in Hongyuan County, where about 85 km² of the swampland was converted to grassland by artificial ditches. This was one of the areas where the Red Army encountered huge difficulties in the summer of 1935 during the "Long March". Today, the inter-connected artificial channels have drained about 85% of this area. Surface water accumulation is now restricted to the middle part of the swamp during the rainy season.

Analysis of recent Google Earth imagery (September 20, 2010) indicated that the total area controlled by the artificial ditches in Hongyuan, Ruoergai, and Maqu counties was approximately 1,109, 885, and 834 km², respectively. Two types of swamp drainage exist in these three counties: (1) areas that are drained only by artificial ditches, such as the Riganqiao Swamp and some small pieces of closed Swamp (220 km^2), and (2) areas where artificial ditches were created to extend the natural channel networks to drain the Swamp, such as the Maiwa and Sedi Swamp in Hongyuan County; Reerdaba, Haqiucuogan, and Hahemuqiao Swampin Ruoergai County; Awancang in Maqu County, etc. (429 km^2). The two types of drained areas account for 4.8 and 9.3% of the total area of 4,600 km^2, respectively. These data indicate that the artificial ditches account for around 14.1% of the loss of the Ruoergai Swamp (4,600 km^2).

The total area of drained swamplands in the Ruoergai Swamp which has been impacted by artificial ditches is about 648 km^2. This accounts for around 27% of the total shrinkage area of 2,400 km^2. The artificial ditches drain the areas along both banks of the channel, dehydrating and hardening the peat. Connecting the artificial ditches to natural channels extends the drainage network, encouraging the shrinkage of the swamp while increasing the rates of flow and sediment flux.

4.2.3.4 Riverbed incision and headcut erosion

Headcut erosion plays an important role in draining the Ruoergai Swamp. Field investigations have indicated that the Bai and Hei rivers and their tributaries are incising due to base level changes along the upper Yellow River. The groundwater levels have decreased and drainage network expansion is evident along both sides of the rivers. Incised river banks along the upper Yellow River at the Maqu section are more than 10 m high. Incised river banks are more than 4 m high along the middle Hei River and around 0.5–3.0 m high along the tributaries of the upper Hei River (Fig. 4.13). Incision and channel expansion are evident along one tributary of the lower Hei River. Upstream areas in Hei River are yet to cut through the surface peat layer, with incised channels around 0.5–1.5 m depth. The middle reaches of Hei River have a depth of around 2 m, with the width of the draining belt extending between 10–30 m. Finally, the downstream section of Hei River is around 5 m deep, extending over a width of more than 500 m.

The rates of headcut erosion were estimated by comparing the Google Earth images from September 20, 2010 and the images from June 15, 2013. Although accurate determination of the incision rate was not achieved due to the lack of historical incision depth data, estimates were still made by measuring the channel extension rates relative to the position of the bridges and culverts along Provincial Highway S301 which runs through the Ruoergai Swamp (the highway construction was completed on October 15, 2010). The incision rates were determined for locations at 13 bridges and culverts over the last 3 years (Table 4.1). The estimated average channel incision rate at these locations was 0.24 myr^{-1}. These incision rates are considered to be higher than the actual incision rates within the channel, as there is no peat layer to protect the river bed at these locations, and scouring is likely to be accentuated adjacent to the bridges and culverts.

The rate of the headcut erosion is a critical contributing factor for the swamp shrinkage. Figure 4.14 shows a typical example of the headcut erosion and expansion of the gully network between September 20, 2010 and June 15, 2013. The average rate

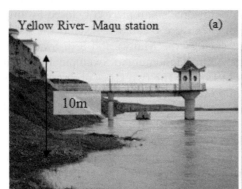

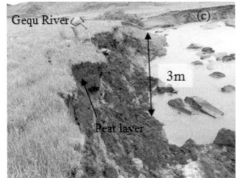

Figure 4.13 Bank erosion at Maqu on the Yellow River, middle Hei River, Gequ River, and Haqu River.

Table 4.1 Headcut incision rate measured at bridges and culverts along Highway S301.

No.	Latitude/Longitude	Net height (m)	Width (m)	Incised depth (m)	Water depth (m)	Incision rate (myr⁻¹)
B-1	N32°57.143′ E103°02.588′	2.65	3.5	0.75	0.00	0.25
B-2	N32°57.143′ E103°02.589′	2.40	5.5	0.65	0.59	0.22
B-3	N33°05.766′ E102°39.059′	2.55	2.0	0.20	0.05	0.07
B-4	N33°05.906′ E102°43.291′	3.00	8.0	0.20	0.48	0.07
B-5	N33°05.607′ E102°46.599′	1.20	2.0	0.85	0.10	0.28
B-6	N33°05.348′ E102°48.272′	1.50	1.5	1.30	0.04	0.43
B-7	N33°05.309′ E102°49.443′	1.20	1.4	0.80	0.05	0.27
B-8	N33°04.659′ E102°50.210′	3.00	2.0	2.70	0.03	0.90
B-9	N33°04.753′ E102°50.574′	5.00	2.0	0.10	0.26	0.03
B-10	N33°04.129′ E102°52.205′	1.00	2.0	0.80	0.05	0.27
B-11	N33°02.168′ E102°55.503′	1.00	0.8	0.50	0.04	0.17
B-12	N32°58.140′ E103°05.045′	2.40	6.0	0.20	0.15	0.07
B-13	N32°57.313′ E103°07.168′	4.50	5.0	0.30	0.20	0.10

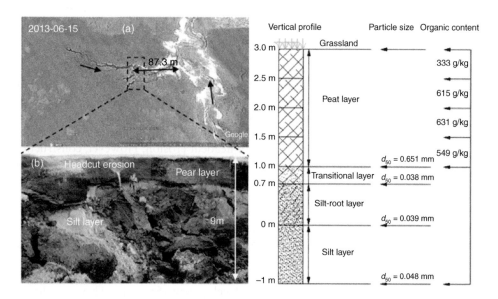

Figure 4.14 (a) A typical headcut erosion point within the Ruoergai Swamp (33°5′6.7″N, 102°50′10.6″E), (b) Measurement of the headcut erosion rate, and (c) Vertical profile of the peat layer.

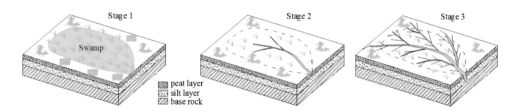

Figure 4.15 Three stages of channel incision and headcut erosion and shrinkage of the Ruoergai Swamp.

of the measured headcut erosion at 30 sites from 2010–2013 was 1.52 m/yr (ranging from 0.68–17.12 m/yr) using Google Earth images.

Channel network expansion of the Bai and Hei tributary systems has progressively drained the Ruoergai Swamp and increased the area of the desiccation belt. Although the peat layer is resistant to erosion, the underlying fine sand and silt layers can be rapidly eroded once the peat layer is breached. Erosion of the sand and silt layers promotes undercutting, bank collapsing and accentuated rates of headcut erosion. A typical 9-m-high hydraulic drop at a headcut site is shown in Figure 4.14(a) and (b) (images from June 15, 2013). In summary, the headcut erosion of 'natural' channels of the tributary systems exerts a significant influence upon the erosion and drainage of a 'closed' swamp.

The peat layer in the Ruoergai Swamp is typically 0.1–3.0 m thick with 442 major peat patches. The total area of the peat land is 4,605 km² (Sun, 1992). Figure 4.14(c) shows the vertical profile (4.0 m depth) of the peat layer at one location of the swamp with the particle median diameter d_{50} of four sediment samples and the organic content of four peat samples. The depth of the pure peat layer is 2.0 m. The materials beneath the peat layer are much more erodible. The median diameter, d_{50}, of the sediment at the boundary of the peat layer and underlying silt is 0.651 mm. Beneath the boundary is a 0.3 m deep transitional layer which primarily consists of peat with some silt. A 0.7-m-deep silt-root layer containing a small quantity of roots lies beneath the transitional layer. A silt sediment unit (median diameters, d_{50}, of 0.038–0.048 mm) is at the bottom. The organic content of the peat layer reflects the content of dead plant roots, varying from 333 to 631 g/kg with depth. These densely interwoven dead grass roots of the peat layer slow down the rate of headcut erosion. However, construction of artificial drainage ditches has lowered the base level and accentuated incision of the tributary systems.

Figure 4.15 shows a schematic representation of the headcut incision of a headcut and subsequent channel expansion which lowers the groundwater levels and accelerates dehydration of swamplands (Safran et al., 2005; Schumm et al., 1984; Tooth et al., 2002). Once the incision breaks through the upper peat layer, rapid undercutting, headward extension, and channel expansion ensue (Fig. 4.15).

During Stage 1, the pre-existing swamp had substantial standing water, and developed under conditions that favoured peat development (Stage 1). These conditions prevailed for many thousands of years. Recent changes in the base level as results from either the construction of artificial ditches or incisions along the trunk stream have modified the relation between different geomorphic processes. During Stage 2, headward channel extension and incision were inhibited by the surficial peat layer during early stages of the changes with the streams. However, once the surficial peat layer was breached, channel incision and expansion via headcut retreat occurred rapidly, causing extensive degradation of the Swamp (Stage 3).

4.3   CRESCENT LAKES IN THE MADUO DESERT

4.3.1   Crescent lakes

Desertification lands in the Yellow River source have a total area of approximately 2,500 km². The majority of the desertification lands are located in the upper watershed of the Youerqu River, which flows into the Yellow River. These desertification lands reside in areas such as the Yellow River valley between northwest Gangnagemacuo and the Huanghe Village, the Yellow River wide valley at northeastern Rigecauchama, the piedmont alluvial fan at the confluence of the Duo and Yellow rivers, the valley of the Kariqu River, the piedmont alluvial fan around the Star Sea, and the wide valley of the north Yueguzonglie River. Analyses of remote sensing images and field survey showed that the crescent lakes in the desert in the Yellow River source are mainly distributed from the Dongqi River mouth to the Eling Lake. The desert in the Yellow River source was chosen as the study area. Multi-period, multi-type, and multi-spectral remote sensing images during 1977–2013 were collected as the base data. In order to make sure that results from the remote sensing analysis were comparable to results from other analysis methods, remote sensing images of different years in the same season were used for the analysis. Meanwhile, field investigations and observations were

1-Hajiangyanchi, 2-Longrecuo, 3-Aronggongmacuo, 4-Arongwamacuo

5-Ayonggamacuo, 6-Gangnagemacuo, 7-Dongcaolong, 8-Rigecuochama

Figure 4.16 Elevation and location of the Maduo desert in the Yellow River source.

conducted for a consecutive three-year period (2011–2013) to study the formation of crescent springs and their movements and ecological importance.

Figure 4.16 shows the location of the desert in the Yellow River source below the Eling Lake. The Eling Lake has an irregular triangular shape. The Yellow River enters the Eling Lake from the left corner (i.e. the east corner), and exists the lake at the upper corner (i.e. the north corner). In the ancient time, the Yellow River flows into the lake at the right corner (i.e. the west corner). Below the Eling Lake, the desert has a banded distribution. The total area of the desert is more than 1,000 km². Its west border connects with the right corner (looking downstream) of the Eling Lake. The entire desert follows the ancient Yellow River course when it was flowing out of the right corner of the Eling Lake. The desert can connect with the Eling Lake by underground seepage because of a layer of gravel and sand which had deposited below the ancient river channel. In 2002, the Huangheyuan Dam started impounding water, the water level in the Eling Lake increased and resulted in the increase of the groundwater level around the lake.

There are more 8,000 crescent lakes in the desert, the vast majority of which are generally stable, but the new crescent springs can move with dune motion. These moving crescent springs are called active crescent springs. In generally, these active crescent lakes gradually stabilize after maintaining water for several years and vegetation has developed on surrounding dunes.

Figure 4.17 shows the measurements of an active and newborn crescent lake. The stable period for an active crescent lake is usually short and the lake stays active.

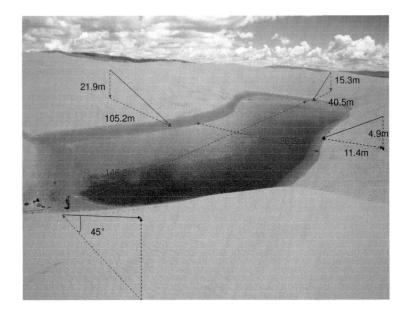

Figure 4.17 Shape measurements of a crescent lake.

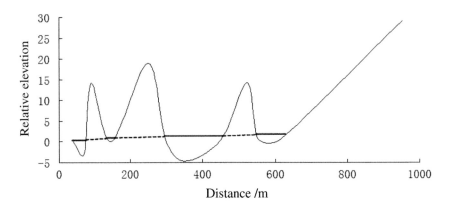

Figure 4.18 Relative water surface elevation difference of four active crescent lakes.

The crescent lake as shown on Figure 4.17 is surrounded by three small dunes. The lake has a width of only 36 m in the direction of the wind and a width of 146 m perpendicular to the wind direction. The upper boundary of the crescent lake is the leeward slope of a sand dune, the slope of which is 25–45°. The opposite side of the lake is the windward slope of another sand dune with a slope less than 15°. The water surface elevation of four nearby crescent lakes was measured by GPS to be approximately 4,230 m. Figure 4.18 shows that the relative water surface elevation difference between the four crescent lakes is within 0.5 m. The water surface elevations were measured by a Laser Ranging Finder with 1 cm accuracy. The small difference in the water levels proves that

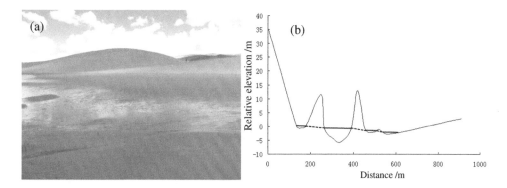

Figure 4.19 (a) Ground water seeping out of dunes and flowing into the depression area due to groundwater rising; and (b) Longitudinal profile of sand and water surfaces along the direction of the largest gradient of the water surface.

the groundwater of the four crescent lakes is connected with each other, even though they are separated by sand dunes hundreds of meters apart.

The active crescent lakes move with the motion of sand dunes. Two years of field observation on several new crescent lakes and the analysis of remote sensing images showed that the rate of a crescent lake movement was 4–10 m/yr. During the 8–9-month-long frozen season, the sand near the downwind side of the crescent lake is blown away. After the thaw, water flows into the depressions created by wind. Meanwhile, the sand is blown into the crescent lake by wind on the other side. In this way, the crescent lake moves with the motion of the sand dunes on both sides (Wang and Han, 2015).

4.3.2 Formation mechanisms of crescent lakes

The Maduo desert in the Yellow River source has developed along the ancient river course, where the groundwater level is relatively high. The groundwater level has increased due to increased snowmelt (as a result of climate change) or other reasons. As a result, the groundwater seeps into the depressions between the sand dunes and forms crescent lakes. Figure 4.19(a) shows the groundwater seeping out of the high dunes and gradually flowing into a depression area. Figure 4.19(b) shows the longitudinal profile of the sand and water surfaces along the direction of the largest gradient of the water surface, of which the left is in the direction of the Eling Lake and the right is in the direction of the Yellow River. Vegetation has started to develop around the crescent lakes in the depressions on the right side of the sand dunes because certain water depth has been maintained in the crescent lakes for a long period of time. Scirpus grows as a dominant species of aquatic vegetation in these crescent lakes, forming a wetland area. The water shown in Figure 4.19(a) is the seepage flow through the sand dunes from the crescent lakes. This wetland has an outlet into the Yellow River with an outflow discharge of 0.1 m³/s. This discharge requires a certain water head, and that is why the water level difference between the crescent lakes and the Yellow River is larger than that between the adjacent crescent lakes.

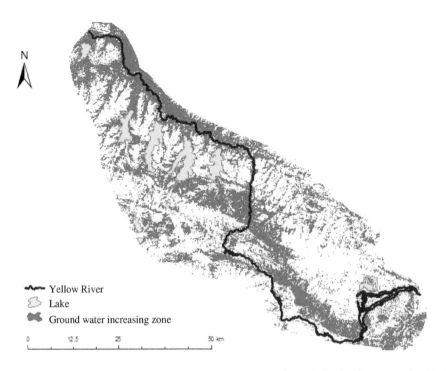

N

Yellow River
Lake
Ground water increasing zone

0   12.5   25   50 km

Figure 4.20 Spatial distribution of ground water rising zones from 2000–2010 around the Maduo Desert downstream from the Eling Lake.

The water level in the Eling Lake rose up to elevation 4,273 m after the impoundment of the Huangheyuan Dam in 2002. Consequently, the groundwater level rose obviously in the desert downstream of the Eling Lake. Remote sensing images of 2001 and 2010 were used to calculate the average radiant intensity which was used to derive the average ground reflectance of 2, 3, 4, and 5 bands. Then the relation between the ground reflectance and the soil moisture content was established and used to calculate the soil moisture percentage. Finally, the model Groundwater Level Distribution Using Remote Sensing (GLDRS) was used to calculate the groundwater level in this area. Figure 4.20 shows the spatial distribution of the groundwater rising zones from 2000–2010 on both sides of the Yellow River downstream from the Eling Lake, around the Xingxinghai Lake, and in the desert of the Yellow River source. The most obvious zone of groundwater rising almost coincides with the desert of the Yellow River source. This is because the groundwater seepage is very active in the ancient course of the Yellow River. Calculation shows that the groundwater level in this region has risen 1–3 m.

The impoundment of the Huangheyuan reservoir and the increase of the water stage in the Eling Lake resulted in the increase of water stages to some extent in adjacent lakes. Figure 4.21 shows the changes of the water surface areas of surrounding eight lakes before and after the impoundment of the Huangheyuan reservoir. The eight lakes had been shrinking before the impoundment in 2002. Especially the Ayong

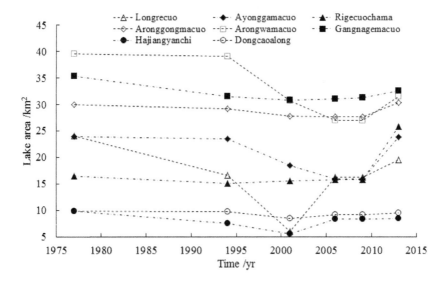

Figure 4.21 Changes of water surface areas of eight lakes surrounding the Eling Lake before and after the impoundment of the Huangheyuan Dam in 2002.

Gamacuo, Longrecuo, Ayong Wamacuo, and Hajiangyanchi lakes had experienced rapid shrinkage during the 20 years before the dam was built. These lakes all stopped shrinking and their water surface areas have been gradually expanding since 2002. The areas of the Ayong Gamacuo, Longregcuo, and Hajiangyanchi lakes have gone back to the values in the 1970s. None of these lakes is connected with the Huangheyuan Dam or the Eling Lake through surface water, which obviously indicates that of the changes in the water surface areas were results of groundwater connections between the lakes.

The number of crescent lakes increased rapidly after the impoundment of the dam, coinciding with the surface area expansion. Some crescent lakes near the Eling Lake have become interconnected and formed large lakes, where only the tops of the sand dunes are exposed on the water surface. Figure 4.22 shows interconnected crescent lakes forming winding water surface and exposed sand dunes. These crescent lakes have formed over a long period of time, and sparse herbaceous vegetation has developed on the surface of the dunes.

Figure 4.23 shows the relative water surface elevation of several new crescent lakes. The average water surface gradient of the crescent lakes is approximately 0.5–1 m/km (Wang and Han, 2016), which coincides with the fact that the water level in the Eling Land is approximately 40 m higher than that in the downstream reach 50 km away.

4.3.3 Ecological effect of crescent lakes

If crescent lakes are recharged by groundwater over a long period of time and the water levels in the lakes are maintained, the activities of surrounding sand dunes will be greatly reduced and vegetation will start to grow on the surface of the

Figure 4.22 Interconnected crescent lakes forming winding water surface and exposed sand dunes.

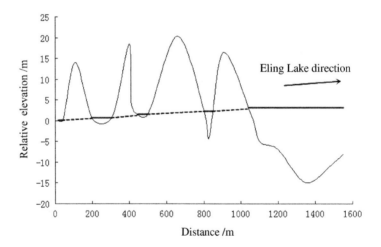

Figure 4.23 Water surface gradient of the new crescent lakes.

desert. The pioneer species which will grow in the desert is *Kobresia Capillifolia*, which covers 2–5% of the total area of the desert. Then sparse vegetation will develop with main species including *Tibet Wormwood, Oxytropis, Alpine Bluegrass, Achnatherum Inebrians, and Saussurea japonica* (covering 10–20% of the total area of the desert). After 5–10 years, the primary vegetation species will include *Lancea Tibetica, Caprifoliaceae, Carex moorcroftii, Aster Asteroides, Astragalus galactites, Kochia scoparia, Anaphalis, Siberia Polygonum, Salsola collina,* and *Potentilla Chinensis.* Meanwhile, *Saussurea Japonica* and *Tibet Wormwood* will still be important species

Figure 4.24 (a) Vegetation cover after the initial stabilization of by crescent lake; and (b) Complex vegetation coverata crescent lake after a long stable period.

(vegetation coverage of 30–50%). When the vegetation coverage reaches over 60%, the vegetation species become even more abundant. *Carex Moorcroftii, Elymus Nutans, Potentilla Chinensis, Virgate Wormwood, Artemisia Frigida, Northwest Stipa, Jujube Soybean, Saussurea Japonica, Edelweiss, Songpan Astragalus* will become the primary species. After a long stable period, complex shrub and grass vegetation will develop with Alpine Spiraea as the dominate shrub. Figure 4.24(a) shows the vegetation cover after the initial stabilization of a new crescent lake. Figure 4.24(b) shows complex vegetation cover at a crescent lake after a long stable period.

The aquatic ecological state of crescent lakes also changes constantly. Aquatic animals colonize in a crescent lake after the first year of its formation. At the early stage, *Gammaridea* is the predominant species. Other species such as *Temoridae, Dytiscidae* and *Corixidae* also exist. The species become more abundant as the crescent lake stays stable for a longer period of time. After a while, *Gammaridea, Radix*, and *Chironomus* become the dominant species, while *Nematoda, Temoridae, Dytiscidae* and *Corixidae* are still important species. Some other species also appear, such as *Planorbidae, Tubificid, Bothrioneurum, Oithonidae, Lepidostomatidae, Psychomyiidae, Cladocera, Rhantus*, and *Noterus*. Basically, the number of species increases with the stable time of the crescent lakes.

The phenomenon of crescent lakes has indicated that the desertification problem with plateaus can likely be solved by building dams to raise the groundwater levels. In fact, many ecological problems in the Yellow River source have been caused by human activities such as dredging and constructing drainage channels. As more and more human activities happen in the Yellow River source, the original plateau marsh, riparian wetland, and plateau meadow have been gradually dewatered as a result of expanding animal husbandry, highway construction and mine industry. A large amount of wind sand are released inadvertently during this process. The accumulation of wind sand causes desertification. Development and growth should be balanced with ecological restoration. Considerations should be given during the planning phase of a hydropower dam to raise the groundwater levels and convert deserts into crescent lakes which can slow down the activities of deserts and restore the ecological conditions. It is feasible to reduce desertification problems on plateaus by encouraging the formation of crescent lakes, however, this approach is not necessarily effective for the deserts in arid areas. The reason is that the temperature and evaporation rates are both low on a plateau, and the groundwater recharges faster than the moisture loss through evaporation. However, the evaporation rates in the deserts in arid areas are very high, which limits the groundwater recharging and makes it difficult to form a sufficient number and size of crescent lakes.

4.4 WETLAND PROTECTION AND RESTORATION

Climate warming and human activities have caused various changes to all wetland types in the Yellow River source. As human activities continue to increase in the Yellow River source, the dynamic changes of the wetlands will continue to intensify. As discussed in the previous sections, the main reasons for the severe shrinkage of the Ruoergai Swamp and plateau meadow wetlands are increased drainage and incision of rivers and ditches within the wetlands. Therefore, the basic strategy to control wetland shrinkage is to control drainage. Pasture expansion by draining wetlands must be stopped as soon as possible. Road and highway constructions should avoid the Swamp by raising the road elevation instead of draining the Swamp. Meanwhile, rock weirs can be constructed at major drainage outlets to control the drainage discharge and slow down water loss in the Swamp. Dam construction (e.g. the Huangheyuan Dam) can increase the number of lakes on plateaus, raise the groundwater level, and encourage the formation of crescent spring wetlands (e.g. Maduo County). Constructing a high dam on the Yellow River valley downstream from Maqu County can also be considered as an alternative strategy to restore the Ruoergai Swamp.

In order to protect and restore the Ruoergai Swamp, effective measures should be taken, e.g. establishing natural reserves, stopping artificial drainage, controlling river incision, prohibiting peat mining, and reducing animal husbandry. In 1998, the Chinese government approved to establish the Ruoergai Swamp Natural Reserve, which had a total area of 1,665 km^2 and was 47 km wide from east to west and 63 km long from north to south. The Swamp Natural Reserve has been listed in the "International important wetland" since 2008. The swamp, meadows, and marsh vegetation in the natural reserve provide very important breeding and habitat environment for local wildlife. Unfortunately, the local climate warming is out of people's control, moreover, human activities (e.g. artificial drainage, peat mining, groundwater exploitation, over-grazing and tourism development, etc.) continue to increase. Although the Ruoergai Swamp Natural Reserve has been established, the shrinkage of the swamp has not stopped especially since headcut erosion and artificial drainage have not been effectively controlled. These unfavorable conditions have imposed adverse effects on the protection of the entire Ruoergai Swamp as well as the maintenance of the natural reserve.

Constructing low-head dams is considered to be an effective approach for wetland restoration. The Flower Lake in the Ruoergai Swamp, which had an original surface water area of 156 ha, had shrunk to a third of its original area by 2011, and desertification had occurred around the lake. A low-head dam was built downstream of the Flower Lake in 2011. Since the dam started impounding water, the Flower Lake quickly expanded to 2,000 ha. The restored Flower Lake looks like a stunning green diamond on the northwest border of Sichuan Province. It has become one of three major wetlands for tourists in China.

Recently, scientists proposed to build a dam 100 km downstream of Maqu County to raise the groundwater level in the Ruoergai Basin and create a new Ruoergai Lake (Xiang and Sheng, 2005). This approach can stop the swamp shrinkage and land desertification in the Ruoergai Basin. Other new strategies have been proposed: Combine close management in regulations, policies, publicity and administration with end-of-pipe treatment (i.e. grass protection and desertification management) to protect the Ruoergai wetland; and control swamp degradation through population control and ecological migration, prohibition of peat mining, animal husbandry and tourism development restrictions (Gao et al., 2006). For restoration of the Riganqiao Swamp it is suggested to reduce and prohibit animal husbandry and tourism development, block artificial ditches and restore vegetation.

REFERENCES

Bai, J.H., Lu, Q.Q., Wang, J.J., Zhao, Q.Q., Ouyang, H., Deng, W., Li, A.N., 2013. Landscape pattern evolution processes of alpine wetlands and their driving factors in the Zoige Plateau of China. Journal of Mountain Science 10, 54–67.

Baidu Encyclopaedia. 2014. Chinese Red Army across the grassland. http://baike.baidu.com/subview/180650/11248854.htm?fr=aladdin (in Chinese).

Barbier, E.B., Acreman, M., Knowler, D., 1997. Economic valuation of wetlands: A guide for policy makers and planners. Ramsar Convention Bureau, Gland, Switzerland.

Boulton, A.J., 2007. Hyporheic rehabilitation in rivers: Restoring vertical connectivity. Freshwater Biology 52(4), 632–650.

Bracken, L.J., Turnbull, L., Wainwright, J., Bogaart, P., 2014. Sediment connectivity: a framework for understanding sediment transfer at multiple scales. Earth Surface Processes and Landforms. doi:10.1002/esp.3635

Brierley, G., Fryirs, K., Jain, V., 2006. Landscape connectivity: the geographic basis of geomorphic applications. Area 38, 165–174.

Brinson, M., Malvarez, A.I., 2002. Temperate freshwater wetlands: Types, status and threats. Environment Conservation 29(2), 115–133.

Cao, L., Fox, A. D., 2009. Birds and people both depend on China's wetlands. Nature 460:173.

Chen, F.H., Bloemendal, J., Zhang, P.Z., Liu, G.X., 1999. An 800 kyr proxy record of climate from lake sediments of the Zoige Basin, eastern Tibetan Plateau. Palaeogeography, Palaeoclimatology, Palaeoecology 151(4), 307–320.

Craddock W.H., Kirby E., Harkins N.W., et al., 2010. Rapid fluvial incision along the Yellow River during headward basin integration. Nature geoscience, 3(3), 209–213.

Craddock, W.H., Kirby, E., Harkins, N.W., Zhang, H.P., Shi, X.H., Liu, J.H., 2010. Rapid fluvial incision along the Yellow River during headward basin integration. Nature Geoscience 3, 209–213.

DeMets, C., Gordon, R.G., Argus, D.F., Stein, S., 1994. Effect of recent revisions to the geomagnetic reversal time scale on estimates of current plate motions. Geophysical Research Letters 21(20), 2191–2194.

Dong, Z., Hu, G.Y., Yan, C.Z., Wang, W.L., Lu, J.F., 2010. Aeolian desertification and its causes in the Zoige Plateau of China's Qinghai-Tibetan Plateau. Environmental Earth Science 59, 1731–1740.

Fryirs, K., 2013. Connectivity in catchment sediment cascades: a fresh look at the sediment delivery problem. Earth Surface Processes and Landforms 38(1), 30–46.

Fryirs, K., 2002. Antecedent landscape controls on river character, behaviour and evolution at the base of the escarpment in Bega catchment, South Coast, New South Wales, Australia. Zeitshrift fur Geomorphologie 46, 475–504.

Fu, B.H., Awata, Y., 2007. Displacement and timing of left-lateral faulting in the Kunlun Fault Zone, northern Tibet, inferred from geologic and geomorphic features. Journal of Asian Earth Sciences 29(2–3), 253–265.

Gao, J., 2006. Degradation factor analysis and solutions of Ruoergai wetland in Sichuan. Sichuan Environment, 25(4), 48–52 (in Chinese).

Harkins, N.W., Kirby, E., Heimsath, A., et al. 2007. Transient fluvial incision in the headwaters of the Yellow River, northeastern Tibet, China. Journal of Geophysical Research, 112(F3), F03S04, doi:10.1029/2006JF000570.

Harkins, N., Kirby, E., Heimsathm A., Robinson, R., Reiser, U., 2007. Transient fluvial incision in the headwaters of the Yellow River, northeastern Tibet, China. Journal of Geophysical Research 112(F3), F03S04, doi:10.1029/2006JF000570.

Hu, Y.R., Maskey, S., Uhlenbrook, S., 2012. Trends in temperature and rainfall extremes in the Yellow River source region, China. Climatic Change 110, 403–429.

Huntley, D.J., Godfrey-Smith, D.I., Thewalt, M.L.W., 1985. Optical dating of sediments. Nature, 313, 105–107.

Li Y., Zhao. Y, Gao S.J., et al., 2011. The peatland area change in past 20 years in the Zoige Basin, eastern Tibetan Plateau. Front. Earth Sci., 5(3), 271–275.

Li, J.J., Fang, X.M., Van der Voo, R., Zhu, J.J., MacNiocaill, C., Ono, Y., Pan, B.T., Zhong, W., Wang, J.L., Sasaki, T., Zhang, Y.T., Cao, J.X., Kang, S.C., Wang, J.M., 1997. Magnetostratigraphic dating of river terraces: Rapid and intermittent incision by the Yellow River of the northeastern margin of the Tibetan Plateau during the Quaternary. Journal of Geophysical Research 102, 10121–10132.

Li, Y., Yan, Z., Gao, S.J., Sun, J.H., Li. F.R., 2011. The peatland area change in past 20 years in the Ruoergai Basin, eastern Tibetan Plateau. Frontiers Earth Science 5(3), 271–275.

Li, Z.W., Wang, Z.Y., Pan, B.Z., Du, J., Brierley, G. J., Yu, G.A., Blue, B., 2013. Analysis of controls upon channel planform at the first great bend of the upper Yellow River, Qinghai-Tibet Plateau. Journal of Geographical Sciences 23(5), 833–848.

Liu, H. Y., Bai, Y. F., 2006. Changing process and mechanism of wetland resources in Ruoergai Plateau, China. Journal of Natural Resources 21(5), 810–818 (in Chinese).

Nicoll, T., Brierley, G. J., Yu, G.A., 2013. A broad overview of landscape diversity of the Yellow River source zone. Journal of Geographical Sciences 23(5), 793–816.

Ning, H.P., Li, G.J., Wang, J.B., 2011. Change features of pan evaporation in Maqu area in the upper reaches of Yellow River in recent 40 years. Journal of Arid Land Resources and Environment 25(8), 113–117 (in Chinese).

Niu, Z. G., Zhang, H.Y., Gong, P., 2011. More protection for China's wetlands. Nature 471, 305.

Niu, Z.G., Zhang, H.Y., Wang, X.W., et al., 2012. Mapping wetland changes in China between 1978 and 2008. Chinese Science Bulletin 57(2), 2813–2823.

Perrineau, A., Van der Woerd, J., Gaudemer, Y., Jing, L.Z., Pik, R., Tapponnier, P., Thuizat, R., Zheng, R.Z., 2011. Incision rate of the Yellow River in northeastern Tibet constrained by [10]Be and [26]Al cosmogenic isotope dating fluvial terraces: Implications for catchment evolution and plateau building. Geological Society, London, Special Publications, 353, 189–219.

Qiu, P.F., Wu, N., Luo, P., Wang, Z.Y., Li, M.H., 2009. Analysis of dynamics and driving factors of wetland landscape in Zoige, eastern Qinghai-Tibetan Plateau. Journal of Mountain Science 6(1), 42–55.

Safran, E.B., Bierman, P.R., Aalto, R., Dunne, T., Whipple, K.X., Caffee, M., 2005. Erosion rates driven by channel network incision in the Bolivian Andes. Earth Surface Processes Landforms 30(8), 1007–1024.

Schumm, S.A., Michael, D.H., Chester, C.W., 1984. Incised channels: Morphology, dynamics, and control. Water Resources Publications, Littleton, Colo.

Sedimentological evidence of piracy of fossil Zoige Lake by the Yellow River. Chinese Science Bulletin 40(18), 1539–1544.

Shen, S.P., Wang, J., Yang, M.J., 2003. Principal factors in retrogression of the Zoige Plateau marsh wetland. Journal of Sichuan Geology 23(2), 123–125 (in Chinese).

Sun, G.Y., 1992. A study on the mineral formation law, classification and reserves of the peat in the Ruoergai Plateau. Journal of Natural Resources 7(4), 334–346 (in Chinese).

Tapponnier, P., Xu, Z.Q., Roger, F., Meyer, B., Arnaud, N., Wittlinger, G., Yang, J.S., 2001. Oblique stepwise rise and growth of the Tibet Plateau. Science 294(5547), 1671–1677.

Tooth, S., McCarthy, T.S., Brandt, D., Hancox, P. J., Morris, R., 2002. Geological controls on the formation of alluvial meanders and floodplain wetlands: The example of the Klip River, Eastern Free State, South Africa. Earth Surface Processes Landforms 27(8), 797–815.

Verhoeven, J.T.A., Artheimer, B., Yin, C., Hefting, M.M., 2006. Regional and global concerns over wetlands and water quality. Trends in Ecology and Evolution 21(2), 96–103.

Wang Zhaoyin and Han Luojie, 2016. Mechanisms of crescent spring lakes in the Huangheyuan Desert. Hydro-Science and Engineering, In press (in Chinese).

Wang, F.B., Han, H.Y., Yan, G., Cao, Q.Y., Zhou, W.J., Li. S.F., 1996. The evolution sequence of Paleo-vegetation and paleo-climate in the northeastern Tibetan Plateau. Science in China (Series D) 26(2), 111–117.

Wang, Y.F., Wang, S.M., Xue, B., Ji, L., Wu, J.L., Xia, W.L., Pan, H.X., Zhang, P.Z., Chen, F.H., 1995.

Xiang, S., Guo, R.Q., Wu, N., Sun, S., 2009. Current status and future prospects of Zoige Marsh in eastern Qinghai-Tibet Plateau. Ecological Engineering 35(4), 553–562.

Xiao, D.R., Tian, B., Tian, K., Yang, Y., 2010. Landscape patterns and their changes in Sichuan Ruoergai Wetland National Nature Reserve. Acta Ecologica Sinica 30(1), 27–32.

Yan, Z.L., Wu, N., 2005. Rangeland privatization and its impacts on the Ruoergai wetlands on the Eastern Tibetan Plateau. Journal of Mountain Science 2(2), 105–115.

Yang, Y.X., 1999. Ecological environment deterioration, mire degeneration and their formation mechanism in the Zoige plateau. Journal of Mountain Science 17(4), 318–323 (in Chinese).

Yu, G.A., Brierley, G., Huang, H.Q., Wang, Z., Blue, B., Ma, Y., 2014. An environmental gradient of vegetative controls upon channel planform in the source region of the Yangtze and Yellow Rivers. Catena, 119, 143–153.

Zhang Xiaoyun, Lu Xianguo. Multiple criteria evaluation of ecosystem services for the Ruoergai Plateau Marshes in southwest China. Ecological Economics, 2010, 69, 1463–1470.

Zhang, X.H., Liu H.Y., Baker, C., Graham, S., 2012. Restoration approaches used for degraded peatlands in Ruoergai (Zoige), Tibetan Plateau, China, for sustainable land management. Ecological Engineering 38, 86–92.

Zhang, X.Y., Lu, X.G., 2010. Multiple criteria evaluation of ecosystem services for the Ruoergai Plateau marshes in southwest China. Ecological Economics 69(7), 1463–1470.

Zhou, D.M., Gong, H.L., Wang, Y.Y., Khan, S., Zhao, K.Y., 2009. Driving forces for the marsh wetland degradation in the Honghe National Nature Reserve in Sanjiang Plain, Northeast China. Environmental Model Assessment 14(1), 101–111.

Chapter 5

Desertification and restoration strategies

Desertification is defined as the phenomenon in which aeolian soil erosion happens, causing sand dunes to form and move, and deserts to form due to climate change and human activities. Desertification has become a focus of the world's ecological and environmental problems. In June 2006, the United Nations Environment Programme (UNEP) published a report titled "The Global Desert Outlook" pointing out that global desertification is increasingly severe. Desert area has reached 38,000,000 km² and accounts for 25% of global land area (UNEP, 2006). More seriously, this number is increasing and 50,000–70,000 km² of fertile land has been lost due to desertification every year at the end of the 20th century. The Sahara Desert expands southward every year with an increasing area of 15,000 km² (UNEP, 2006). China is also one of the countries experiencing serious desertification. According to the State Forestry Bureau report, the desert area in China in 1949 was 670,000 km² and in 1985 it reached 1,300,000 km², and then in 2005 it further increased to 1,730,000 km² (SFB, 2005).

Desertification seriously impacts agriculture and animal husbandry. As agricultural land area decreases and land productivity also drops, serious economic losses result. Land productivity decreased an estimated 20–25 percent and pasture yield declined an estimated 30–40 percent (Yang and Zhou, 2000; Wang, 2004). Worldwide desertification has caused agricultural production losses of $26 billion each year (Shi, 1991; Houghton, 1998). In the middle of the last century, areas on the edge of the Sahara Desert in Africa suffered a continuous drought of 17 years, causing huge economic losses and disaster to neighbouring countries, resulting in more than 200 deaths and shocking the world (Ren, 1990). Desertification also has destroyed wildlife habitat. Species extinction caused by desertification has aroused high attention from the scientific community and governments (Zhao, 2005; Yang, 2010; Wang, 2006b). Aeolian sand disasters caused by desertification have an environmental effect on regional and global scales. Agricultural land, rivers, lakes, houses, roads, railways, and reservoirs have been destroyed when the aeolian sand passed through. Roads were diverted, infrastructure was rebuilt, and even villages were forced to move. Desertification has influenced human production and living and has become a serious threat to human life. Sandstorms from the northwest often affect the spring atmospheric visibility in Beijing, and even sometimes have affected the Korean peninsula and Japan (Zhang and Zhao, 2008).

In the past century desertification occurred in many areas of China, especially in the Qinghai-Tibet Plateau. The north part of the plateau is dry and cold. The vegetation is very vulnerable and has low resilience. Overgrazing and agriculture development caused serious desertification in Sanjiangyuyan. This chapter discusses the phenomenon of desertification in Sanjiangyuan and management strategies.

5.1   DESERTIFICATION IN SANJIANGYUAN

5.1.1   Sanjiangyuan and desertification

The source area of the Yangtze, Yellow, and Lancang (Mekong) rivers, situated in the northeastern Qinghai-Tibet Plateau, is named Chinese Sanjiangyuan (source of the three rivers), as shown in Fig. 1.6. Sanjiangyuan provides 25% of the Yangtze River's water, 49% of the Yellow River's water, and 15% of the Lancang River's water, in total supplying 60 billion m^3 water to downstream every year as such it is described as the water tower of China (Dong et al., 2002). Sanjiangyuan has an arctic climate, with an average elevation of 3500 to 4800 m, an annual average temperature of 4.0 to 6.0°C, annual precipitation of 250–350 mm, and annual average wind speed of 3.0 to 5.0 m/s (Hu et al., 2008). In the past several decades, desertification of grass areas in the source area of the Yangtze and Yellow rivers was serious and became a threat to pasture land and ecology. In this chapter, land desertification and restoration strategies for Sanjiangyuan are discussed and a theoretical basis for management of desertification on the plateau is provided.

In the Alpine desertification area of Sanjiangyuan, the soil vertical profile is composed of an interaction and superposition of sedimentary facies including an aeolian sand layer, a loess layer, and paleosol. In the soil surface aeolian sand loess, paleosol, and lacustrine sediment coexist. Since the last glacial period the process of desertification has occurred and expanded and the opposite process that desert areas have shrunk and disappeared have alternately happened in Sanjiangyuan (Tao and Dong, 1994). The desertification process of various time scales was not completed and was replaced by the strong opposite process to reverse desertification. Climate fluctuations are the main factor influencing these processes. Additionally, throughout human history unreasonable economic activities have also intensified the development of desertification (Dong et al., 1987).

Since the 1950s, scientific investigations and remote sensing data analysis have been performed on Sanjinagyuan several times by national experts. It was found that desertification was serious in the unit landforms including intermountain basins, river valleys, river basins, and alluvial plains on the plateau (SFB, 2009). Both the area and degree of desertification have developed towards deterioration (Hu et al., 2008). In the last 40 years, human activities have intensified, especially overgrazing, and rats and rabbit marmots have excessively reproduced and destroyed the prairie, which leads to desertification. Further, climate change caused the annual average temperature to increase 0.7 to 0.8°C and intensified desertification on the Plateau (Wang et al., 1998; Wang et al., 2001). Desertification in Sanjiangyuan has become the most serious current ecological problem (Dai et al., 2010).

Vegetation deterioration is the prelude to desertification. Ancient fluvial deposits gradually dried up together with the Plateau uplift exposing the earth's surface have led to vegetation deterioration. Alpine aeolian sandy soil is widely distributed around

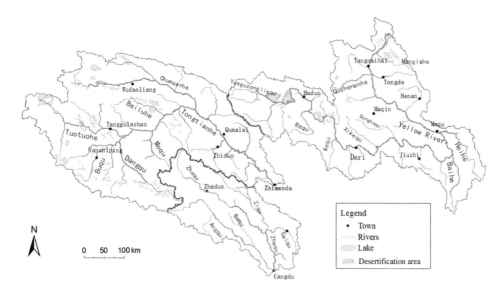

Figure 5.1 Distribution of streams and desertification areas in Sanjiangyuan.

flood deposits in Sanjiangyuan. The deposits were covered with herbaceous vegetation under the influence of cold climate and the plateau geographical environment. In 1990s, modern moving sand dunes, the Pleistocene and the Holocene deserts, and deserts of earlier geological times were stabilized by vegetation and formed the aeolian sandy soil (Fang et al., 2004). Vegetation degradation occurred with part of the vegetation damage resulted from climate change and human activities. Fine sand under the vegetation cover provided abundant solid material for aeolian sand movement (Zhao, 1991). A portion of the fine sand was exposed and transported by strong winds and formed moving sand dunes. The movement of sand dunes buries and destroys herbaceous vegetation, releases more sand, and causes morphological changes (Zhang, 2005; Feng et al., 2004).

Figure 5.1 shows the distribution of desertification areas, streams, and lakes in Sanjiangyuan. This distribution diagram was obtained using a Geographical Information System (ArcGIS) based on the SRTM database and the 1:250,000 topography database from national administration of surveying, mapping and geoinformation. Desertification areas are concentrated in the source area of the Yellow and Yangtze rivers (note: the Tongtianhe is the most upstream reach of the Yangtze River). The total area of the desertification areas is about 3000 km². At present modern aeolian sand movement is very strong and threatens the ecological environment and pasture land (Fig. 5.2). More seriously the plateau has become one of the important sand and dust source areas in China. Due to the special elevation of the "Roof of the World," sand and dust easily enter the westerly jet stream due to strong winds. The plateau has become one of the main sand and dust source areas of the global remote transmission, and this sand and dust even affects the global climate (Han et al., 2004).

(a)   Mangla in Guinan   (b)   Shapotou in Guinan

Figure 5.2 Movement of sand dunes buries and destroys herbaceous vegetation, releases more sand, and causes morphological changes in the source area of the Yellow River.

5.1.2 Characteristics of aeolian sand

A huge amount of aeolian sand, mainly composed of quartz mineral particles and with a size of 0.05 to 0.25 mm, has been stored in the weathering products of rocks. These sand particles are easily transported by wind and moved by saltation on the desert surface, but cannot rise to high altitudes. In the movement process the particles do not move as discrete particles but gather together when moving. Desertification is the result of this aeolian movement of sand particles in groups that buries the vegetation and forms moving sand dunes. Without human interference these particles are stored in different forms. Once the particles are released and gather together, the desert areas form.

Aeolian sand deposits are widely distributed in Sanjiangyuan including moving sand dunes, semi-fixed sand dunes, and fixed sand deposits. About 90 percent of the particles composing the moving sand dunes are aeolian sandy particles. The vegetation coverage of moving sand dunes is less than 5%. The surface sand of a moving dune is easily moved by wind. About 80 percent of the particles composing semi-fixed sand dunes are aeolian sandy particles. The vegetation coverage of semi-fixed sand dunes is more than 5%. Part of a semi-fixed sand dune is covered by a crusting layer or humus layer. The sand under the cover is protected from aeolian sand erosion. Semi-fixed sand dunes remain stable under non-extreme conditions. Once the crusting layer is destroyed, fine sand can rise up and move around. Semi-fixed dunes are flattened to a certain extent, but still maintain the distinctive shape of a sand dune. Only about 70 percent of particles composing fixed sand deposits are aeolian sandy particles. The vegetation cover of a fixed sand deposit is over 60%. In the well-revegetated area the sand deposit has experienced a long-term weathering and humification. A humus layer of a certain depth on the fixed sand deposit effectively stops aeolian sand erosion.

Field investigations of aeolian sand deposits have mainly been performed in the source areas of the Yangtze and Yellow rivers. Study regions include Wudaoliang, Tanggulashan town, and Budongquan in the source areas of the Yangtze River and

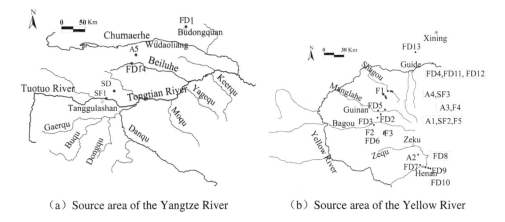

(a) Source area of the Yangtze River   (b) Source area of the Yellow River

Figure 5.3 Locations of sampling sites in Sanjiangyuan.

Guide, Guinan, Zeku, and Henan in the source areas of the Yellow River. Figure 5.3 shows the streams in Sanjiangyuan and sampling sites including 5 moving sand dunes (A1–A5), 3 semi-fixed sand dunes (SF1–SF3), 5 fixed sand deposits (F1–F5), 14 fluvial deposits (FD1–FD14), and 1 slope deposit (SD). Geographical characteristics were measured using Global Positioning System (GPS) receivers with a maximum error of one meter. The height and slope of sand dunes were measured by using a range finder and a compass. Sediment samples of 0.3 kg at the surface layer of each sediment deposit site were taken by a soil sampler. The size distribution of the particles larger than 0.075 mm was obtained by sieving, whereas the size distribution of the particles smaller than 0.075 mm was obtained by a laser particle analyzer (Niu et al., 2006). The mineral composition of the sediment samples was tested by an X-ray diffraction instrument with a stability of 0.01%.

Desertification in Sanjiangyuan results from the release and aggregation of aeolian sand particles stored in the fixed sand deposits and semi-fixed sand dunes. The aeolian sand particles were found to be from fluvial deposits when traced to the original source. Deposits of gravity erosion from hillsides, debris flow, and river sediment have a wide grain size distribution. After wind erosion, the remaining coarse particles in the surface layer protect the fine sand particles from being eroded by wind again. However, these deposits would be sorted after river transportation. Fine sand particles are concentrated in the deposits on the floodplain. Once the wind acts, these fine particles would be moved and transported a certain distance, finally forming moving sand dunes. Figure 5.4 shows the grain size composition curve comparison for moving sand dunes (A1, A3, A4, and A5) and nearby fluvial deposits (FD1, FD2, FD4, and FD5) in the Chumaer River basin in the source area of the Yangtze River and the Shagou River basin and the Mangla river basin in the source area of the Yellow River. The results showed that the particle sizes of aeolian sand deposits were concentrated. That is, particles with size of 0.1–0.2 mm in the fluvial deposits were sorted by wind and formed sand dunes.

Figure 5.5 shows the mineral composition curve comparison for moving sand dunes (A1, A3, A4, and A5) and fluvial deposits (FD1, FD2, FD4, and FD5). The

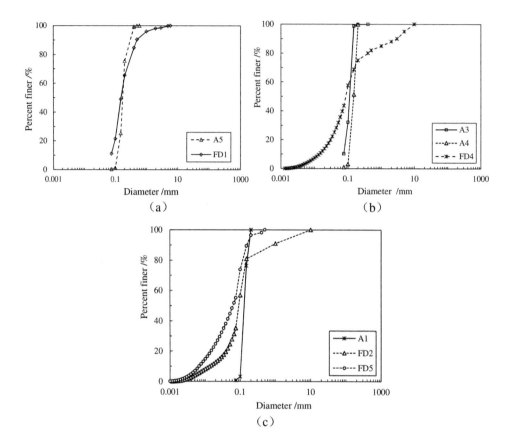

Figure 5.4 Grain size distributions of moving aeulian sand dunes (A1, A3, A4, and A5) and fluvial deposits (FD1, FD2, FD4, and FD5).

results showed that the mineral composition content of fluvial deposits were basically the same as for aeolian sand deposits. The mineral content was mainly composed of Quartz, Potassium feldspar, Albite, Calcspar, Hornblende, and Clay minerals. Clay mineral particles are defined as silicate mineral particles finer than 0.002 mm including kaolinite, illite, smectite, chlorite, etc. The clay mineral content decreased and the quartz particles increased during the process of aeolian sand sorting. It was found that quartz particles of the fluvial deposits enter into aeolian sand, while the clay mineral particles are blown off.

The sediment samples of 13 aeolian sand deposits and 15 fluvial deposits were analyzed and tested for size distributions. The size distributions of 71 sediment samples of moving dunes, water erosion, and gravity erosion in other regions also were collected. Size dis tributions of 21 moving dunes were collected from other regions including valleys on the Plateau in China, arid desert, arid semi-desert, semi-humid regions in China (Chen, 1998), the Sahara Desert in North Africa, and deserts in Saudi Arabia (Abolkhair 1986; Wu 1987; Khatelli et al., 1998; Liu et al., 2005). Size distributions of 59 other deposits including water erosion and gravity erosion from Tibet,

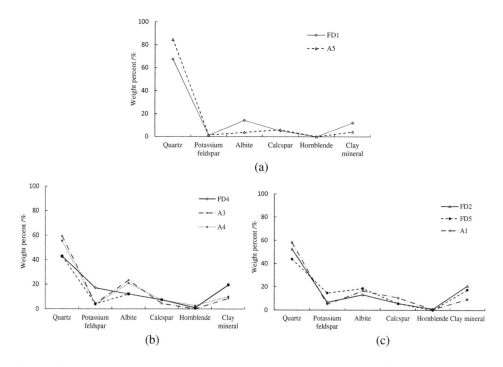

Figure 5.5 Mineral composition comparison for moving sand dunes (A1, A3, A4, and A5) and fluvial deposits (FD1, FD2, FD4, and FD5).

Sichuan, Yunnan, and the Yellow River Delta were collected (He and Liu 2007; Kang et al., 2004; Li et al., 2010; Ye 1992). In particular, water erosion deposits included slope deposits, debris flow deposits (including gully and slope debris flow deposits), and river sediment deposits (including flood deposits, bed load, and bed materials). Gravity erosion deposits included landslide deposits and avalanche deposits. Above all, the size distributions of 46 aeolian sand deposits, 2 slope deposits, 5 landslide deposits, 7 avalanche deposits, 24 debris flow deposits, and 27 river sediment deposits were obtained by field investigations, analytical tests, and the literature.

Figure 5.6 shows the median diameters and sorting coefficients, defined as $C_v = \sqrt{D84/D16}$, of the aeolian sand deposits and of other deposits, where $D84$ is the diameter of a particle for which 84 percent of the sediment is finer and $D16$ is the diameter of a particle for which 16 percent of the sediment is finer. The abscissa is median diameter, $D50$, and the ordinate is the sorting coefficient C_v. The results showed that the size distributions of aeolian sand deposits in different regions were almost the same (see the oval area around $D50 = 0.1$ mm in Fig. 5.6). The grain diameters were in the range of 0.05–0.25 mm and the sorting coefficients were between 1 and 2.5. The sediment of moving sand dunes had an average median diameter of 0.15 mm with an average sorting coefficient of 1.35. The sediment particles within this range are the key material source for the formation of moving sand dunes. Other types of deposits had very different size distributions and much larger sorting coefficients. The median diameters of gravity erosion deposits vary in a great range from 20 to

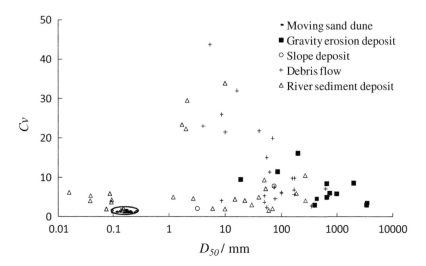

Figure 5.6   Median diameters and sorting coefficients of aeolian sand in comparison with those of other deposits.

5000 mm. The largest sorting coefficient is more than 15. The median diameters of debris flow deposits varied in a great range from 5 to 1000 mm. The largest sorting coefficient was more than 40. The median diameters of the river sediment deposits varied in a great range of 0.01 to 1000 mm. The largest sorting coefficient was more than 30. Therefore, the aeolian sand deposits in different regions had similar size distributions and minimal sorting coefficients, and were totally different from the other sediment deposits. These are the identifying characteristics of aeolian sand.

5.2 VEGETATION DEVELOPMENT ON SAND DUNES

5.2.1 Interaction of fixed sand dunes and vegetation

In some desertification areas where the annual rainfall is more than 200 mm, if sand dunes stay still without movement, vegetation will recover. Therefore, the key for desertification and afforestation is to prevent the movement of sand dunes and the expansion of desertification areas. Through the impacts on wind erosion, aeolian sand transport, and sand dune fixation (Van de Ven et al., 1989; Wolfe and Nickling, 1993; Buckley, 1996), vegetation prevents the movement of sand dunes (Durán and Hernmann, 2006). The vegetation development process can promote the process of sand dune stabilization. Vegetation promotes and accelerates the stabilization of sand dunes and the ecological restoration of desertification areas. Aeolian sand deposits of different types in Maduo were studied and it was found that the organic content, soil nitrogen, soil phosphorus, total number of bacteria, and microbial species increased with time (Lin et al., 2007; Qi et al., 2005).

In this study, the effects of vegetation development on aeolian sand dunes were evaluated. A field investigation of the sampling sites shown in Fig. 5.3 was done, these

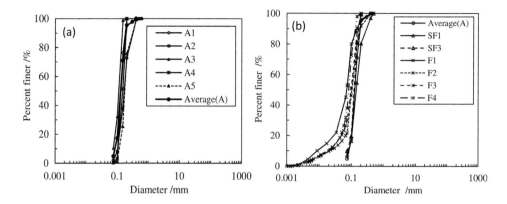

Figure 5.7 Grain size distributions: (a) Moving sand dunes; (b) Semi-fixed sand dunes (SF) and fixed aeolian sand deposits under the action of vegetation.

sites included 5 moving sand dunes (A1–A5), 3 semi-fixed sand dunes (SF1–SF3), and 5 fixed sand deposits (F1–F5). Moving sand dunes are widely distributed on the fluvial plain and mountains in the Sanjiangyuan region. Moving sand dunes have a regular barchan shape, with a height of 7 to 10 m (A1, A2, and A5). Moving sand dunes, A1, A2, and A5, were separately located in different huge desertification areas. Moving sand dunes, A3 and A4, with a height of about 2–3 m were in different small desertification areas. Semi-fixed sand dunes, SF1 and SF3, with vegetation coverage of about 35% have already experienced long-term effects of natural vegetation. Semi-fixed sand dune, SF2, with vegetation coverage of about 20% has already experienced vegetation effects for several years after afforestation. Fixed sand deposits, F1-F4, with vegetation coverage over 65% have already experienced natural vegetation effects for tens or hundreds of years. Fixed sand deposit, F5, with vegetation coverage of about 40% has already experienced vegetation effects for more than ten years after afforestation. Fixed sand deposits are flat without any movement of sand.

Figure 5.7 shows the size distribution curves of moving sand dunes in comparison with those of aeolian sand deposits. The particle size of moving sand dunes was between 0.075–0.2 mm with an average median size of 0.15 mm. The curve Average (A) in Fig. 5.7(a) represents the average cumulative size distributions of moving sand dunes A1-A5. After sand dunes were stabilized by fixing projects or vegetation measures, surface particles in the deposit have been weathered under the action of precipitation and vegetation effects. Figure 5.7(b) shows the size distributions of semi-fixed sand dunes, SF1, and SF3, and fixed sand deposits, F1–F4, under the actions of natural vegetation. The results showed that the fractions of solid particles finer than 0.075 mm in semi-fixed sand dunes and fixed sand deposits were 10% and 20%, respectively. This material included fine sand, powder sand, and clay particles. The fractions of clay particles produced in the soil forming process for moving sand dunes, semi-fixed sand dunes, and fixed sand deposits were less than 5%, 5–10%, and 10–20%, respectively. Due to the increase of fine sand, the sorting coefficient became larger. The average sorting coefficient of moving sand dunes, semi-fixed sand dunes, and fixed sand deposits were 1.35, 1.5, and 2.2, respectively.

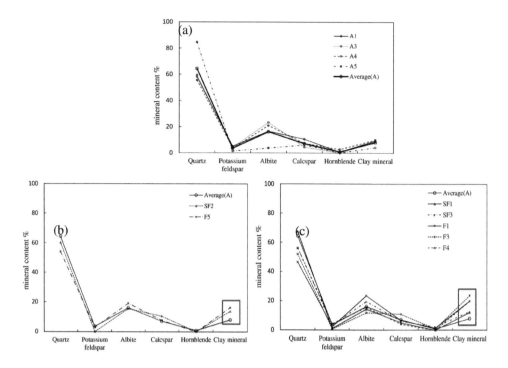

Figure 5.8 Mineral composition of aeolian sand deposits in Sanjiangyuan: (a) Moving dunes (b) Semi-fixed and fixed aeulian sand deposits under the action of natural vegetation; (c) Semi-fixed and fixed aeulian sand deposits under the action of artificial and natural vegetation.

In the process of sand dune stabilization, clay mineral particles are produced due to hydrolyzation and carbonation of feldspar and carbonate mineral particles. In addition, humus particles are produced by organic matter under microbial action. Therefore, based on the analysis of mineral composition and humus content, it is clear whether the sand dune has been stabilized. Figure 5.8 shows the mineral composition and content of sediment samples in Sanjiangyuan. The mineral composition of moving sand dunes is quartz, feldspar, carbonate, hornblende, and clay minerals. The clay mineral particles of account for 8% moving sand dunes (Fig. 5.8(a)). A small portion of non-clay mineral particles (except for quartz) in the deposit has been weathered and changed to clay minerals after sand dunes become fixed. Figure 5.8(b) shows the mineral composition and content of semi-fixed sand dunes and fixed sand deposits under the action of natural vegetation. The fractions of clay mineral particles in the surface layer of fixed sand deposits and semi-fixed sand dunes are 22% and 10%, respectively. Figure 5.8(c) shows the mineral composition and content of semi-fixed sand dunes and fixed sand deposits under the action of artificial and natural vegetation. The fractions of clay mineral particles in the surface layer of fixed sand deposits and semi-fixed sand dunes are 18% and 15%, respectively. In addition, the humus contents of some sediment samples were evaluated. The humus contents of moving sand dunes, A1, A3, and A4, were 4.9 g/kg, 5.23 g/kg, and 4.88 g/kg, respectively. The humus contents of semi-fixed sand dune, SF3, and fixed sand deposit, F1, were 7.92 g/kg and 15.08 g/kg, respectively. In conclusion, the clay mineral particles in fixed sand deposits were twice those in moving sand dunes, and the semi-fixed sand dunes were in between. The content of humus in fixed sand deposits was three times more than that in moving sand dunes.

During the transformation of moving sand dunes into fixed sand deposits, clay mineral particles and humus particles are produced and form the crusting layer and humus layer on the surface of the sand. The crusting layer and humus layer play a key role in protecting the sand dunes from aeolian sand erosion. The crusting layers of semi-fixed sand dunes are several millimeters or centimeters thick. The humus layers of fixed sand deposits are several centimeters or tens of centimeters thick. The fixed sand deposits under the action of artificial and natural vegetation experience a short-term soil forming process, therefore, the humus layer is several centimeters thick. Whereas, the fixed sand deposits under the action of natural vegetation experience a soil forming process of several tens or hundreds of years, therefore, the humus layer is tens of centimeters thick and contains a lot of plant roots. Compared with aeolian sand deposits under the action of natural vegetation, the soil forming process of aeolian sand deposits under the action of artificial and natural vegetation is much faster. The content of fine particles in aeolian sand deposits under the action of artificial and natural vegetation after tens or hundreds of years is equal to that of aeolian sand deposits after tens-of-years of action of natural vegetation. These facts indicate that artificial vegetation promotes and accelerates the sediment particle weathering process.

5.2.2 Vegetation succession in desertification areas

In order to control sandstorm disasters and prevent expansion of desertification, vegetation restoration projects have been performed in the desertification areas in Sanjiangyuan. Vegetation restoration projects include mechanical sand fixation, tree planting, straw checkerboard barriers, gravel sand fixation, forest reservation, among others (Zhang et al., 2003b). In the process of sand dune stabilization, zonal grassland species grow and a vegetation succession process occurs.

The vegetation of moving sand dunes, A1 and A2, semi-fixed sand dunes, SF1 and SF2, and fixed sand deposits, F1 and F5, were investigated. Moving sand dune, A1, semi-fixed sand dune, SF2, and fixed sand deposit, F5, were in the Mugetan desertification area in the source area of the Yellow River. Vegetation of uniform population distribution was measured in three 1 × 1 m quadrats. Vegetation of concentrated population distribution was measured in one 5 × 5 m or 10 × 10 m quadrat and two 1 × 1 m quadrats. Table 5.1 lists the investigation results of the surface vegetation of moving sand dune, A2, semi-fixed sand dune, SF1, and fixed sand deposit, F1, under the action of natural vegetation. The vegetation life structure of semi-fixed sand dune, SF1, and fixed sand deposit, F1, under the action of natural vegetation only contained an herbaceous layer. Table 5.2 shows the investigation results of the surface vegetation of moving sand dune, A1, semi-fixed sand dune, SF2, and fixed sand deposit, F5, under the action of artificial and natural vegetation. The vegetation life structure of moving sand dune, A1, semi-fixed sand dune, SF2, and fixed sand deposit, F5, after afforestation included an herbaceous layer and a shrub layer. The species number and number, height, and coverage of the vegetation on the surface of sand dunes were measured by

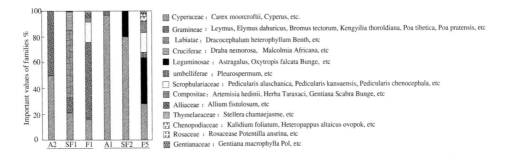

Figure 5.9 Importance values of families composed of herbaceous vegetation of aeolian sand deposits.

appropriate area and number of quadrats. Plant samples were gathered and identified. During the process of moving sand dunes transforming into fixed sand deposits, the vegetation composition and coverage on the windward slope was different from those on the leeward slope on the surface of the sand dune. The coverage, richness, and average vegetation height on the sand dune were calculated by the investigation results of vegetation quadrats. The richness of vegetation is the number of species.

The vegetation characteristic on the windward slope was almost the same as that on the leeward slope of the moving sand dune, semi-fixed sand dune, and fixed sand deposit under the action of natural vegetation. Under the action of natural vegetation, moving sand dune, A2, transformed into semi-fixed sand dune, SF1, and fixed sand deposit, F1. Cyperaceae and Gramineae were the dominant plant families on the surface of moving sand dune A2 as shown in Fig. 5.9. Labiatae, Cyperaceae, Cruciferae, and Gramineae were the dominant plant families on the surface of semi-fixed sand dune SF1 as shown in Fig. 5.9. Gramineae, Scrophulariaceae and Cyperaceae were the dominant plant families on the surface of fixed sand deposit F1 as shown in Fig. 5.9. *Carex moorcroftii* of the Cyperaceae family was the pioneer species for the stability of the sand dunes. After natural vegetation development and succession over tens or hundreds of years, *Poa tibetica* of the Gramineae family became the dominant species, and the vegetation coverage on the fixed sand deposit reached 95%, the species richness became 11, the average vegetation height reached 40 cm (Table 5.1).

After artificial planting on the surface of the sand dune, moving sand dune, A1, transformed into semi-fixed sand dune, SF2, and fixed sand deposit, F5. Cyperaceae is the dominant plant family on the surface of moving sand dune A1 as shown in Fig. 5.9. Cyperaceae and Leguminosae are the dominant plant families on the surface of semi-fixed sand dune SF2 as shown in Fig. 5.9. Cyperaceae, Leguminosae, Cruciferae, and Scrophulariaceae are the dominant pland families on the surface of fixed sand deposit F5 as shown in Fig. 5.9. Under the action of Cathay poplar and salix Mongolia on the surface of sand dunes, herbaceous layer vegetation developed. *Cyperus* and *Carex moorcroftii* of the Cyperaceae family and *Bromus tectorum* of the Gramineae familty are the dominant species (Table 5.2). After vegetation succession of several tens of years, the coverage of the shrub layer vegetation was over 30% and the average vegetation height was 300 cm, but the species richness remained unchanged. While the coverage of the herbaceous layer was more than 60%, the species richness was 11, and

Table 5.1 Vegetation characteristics of sand dunes under the action of natural vegetation.

No.	Height of dune (m)	Slope	Vegetation	Coverage (%)	Richness	Vegetation height (cm)	Species composition
A2	7	Wind-ward and lee-ward	No vegetation	0	0	0	No vegetation
SFI	5	Wind-ward and lee-ward slopes	Herbaceous layer	30	8	20	*Dracocephalum heterophyllum Benth; Carex moorcroftii; Draba nemorosa; Elymus dahuricus; Kengyilia thoroldiana; Astragalus; Malcolmia Africana; Pleurospermum*
FI	0.5	Wind-ward and lee-ward slopes	Herbaceous layer	95	11	40	*Poa tibetica; Elymus dahuricus; Carex moorcroftii; Pedicularis alaschanica; Poa pratensis; Pedicularis kansuensis; Herba Taraxaci; Allium fistulosum; Pedicularis chenocephala; Stellera chamaejasme*

the average vegetation height was 45 cm. Meanwhile, *Oxytropis falcata Bunge* of the Leguminosae family became the dominant species.

Moving sand dune, A2, transformed into fixed sand deposit, F1, after experiencing natural vegetation development and succession over tens or hundreds of years. Moving sand dune, A1, transformed into fixed sand deposit, F5, which experienced vegetation development and succession over tens of years after artificial planting. The number of plant species on the surface of fixed sand deposits, F1 and F5, was equal. The species composition of vegetation on the surface of fixed sand deposit, F1, was different from that of fixed sand deposit, F5. During the process of moving sand dunes transforming to fixed sand deposits, the vegetation coverage and the species richness increased significantly. The vegetation coverage increased from less than 5% to over 60%. The number of species increased from several strains to ten strains. With the development and succession of vegetation, the height of the herbaceous vegetation increased from several centimeters to tens of centimeters, and the height of the shrub vegetation increased from several tens of centimeters to several hundreds of centimeters.

During the process of vegetation development and succession, the importance value of each species changed. The importance value of pioneer species gradually decreased and the importance values of the dominant species of different stages increased. Figure 5.9 shows the importance values of herbaceous plant families on the surface of aeolian sand deposits. The importance value of herbaceous plant families was calculated by accumulating the importance values of herbaceous plant species of

Table 5.2 Vegetation characteristics of sand dunes under the action of artificial and natural vegetation.

No.	Height of dune (m)	Slope type	Vegetation Structure	Coverage (%)	Richness	Vegetation height (cm)	Species composition
A1	9	Wind-ward slope	Shrub layer	0.015	1	60	Populus cathayana
			Herbaceous layer	2.5	1	8	No vegetation
		Lee-ward slope	Shrub layer	0.01	1	60	Populus cathayana
			Herbaceous layer	3	3	10	Cyperus; Carex moorcroftii; Bromus tectorum
SF2	4	Wind-ward slope	Shrub layer	0.09	1	160	Populus cathayana
			Herbaceous layer	3	1	25	Carex moorcroftii
		Lee-ward slope	Herbaceous layer	30	2	25	Carex moorcroftii; Oxytropis falcata Bunge
F5	1	Wind-ward and Lee-ward slopes	Shrub layer	30	2	300	Populus cathayana; Salix psammophila
			Herbaceous layer	45	11	45	Oxytropis falcata Bunge; Carex moorcroftii; Artemisia hedinii; Gentiana Scabra Bunge; Kalidium foliatum; Pedicularis chenocephala; Pleurospermum; Heteropappus altaicus ovopok; Rosacease Potentilla ansrina; Gentiana macrophylla Poll

the same family. The importance value of herbaceous plant species is the sum of the relative number, the relative coverage, and the relative frequency of this species. The relative number is the percentage of the number of individuals of the species among the total number of individuals of all species. The relative coverage is the percentage of the coverage of the species among that of all species. The relative frequency is the percentage of the frequency of the species among that of all species (Fu, 2006).

A few annual herbaceous plants including Cyperaceae and Gramineae families, were sparsely distributed on the surfaces of moving sand dunes and became pioneer species for the stability of sand dunes. With the development of zonal grass species, the importance value of the Cyperaceae family reduced, and Gramineae and Labiatae became the dominant families after the vegetation succession over hundreds or tens of years, and additionally Gramineae and Scrophulariaceae families were involved in the process of vegetation succession. Under the action of Cathay poplar and salix Mongolia on the surface of sand dunes, the herbaceous layer vegetation developed. The pioneer species were from the Cyperaceae and Gramineae families. After vegetation succession over several tens of years, Leguminosae and Scrophulariaceae became the dominant families, and the Compositae, Rosaceae, and Umbelliferae families also were present. Vegetation development and succession accelerated the accumulation of clay mineral and humus particles, whereas the production of clay mineral particles and crust particles promoted the process of vegetation development and succession. Qingzangtaicao of the Cyperaceae family was the pioneer plant on fixed sand deposits, F1 and F5. F1 and F5 experienced natural vegetation succession over tens or hundreds of years and artificial and natural vegetation effects over a few years, respectively. At present, the vegetation coverage on F1 and F5 are over 70% and the number of species on F1 and F5 is 11. The humus layer of F1 and F5 has a certain thickness, and the fractions of fine sand and clay mineral particles are 10% and 20%, respectively. F1 and F5 have achieved a stable state. Consequently, whether through natural vegetation succession or under the action of artificial and natural vegetation development, once the vegetation of sand dunes develops to a certain degree and the particles in the surface layer have weathered to a certain degree, the aeolian sand deposits remain stable under non-extreme conditions. Artificial afforestation accelerates the development and succession of vegetation and the soil forming process, and promotes the stability of the sand dune and the ecological restoration of the desertification land.

5.3   VEGETATION-AEOLIAN EROSION DYNAMICS

5.3.1   Interaction of aeolian sand dune and vegetation development

Aeolian sand activity affects vegetation growth and productivity and is one of the ecological factors limiting vegetation development (Bendali et al., 1990; Liu et al., 2002). The processes of wind erosion, sand transport, and sand accumulation influence vegetation growth and development. Wind erosion damages soil structure, reduces soil fertility, and, thus, affects the growth of vegetation, and even causes plants to wither and die (Liao, 1980; Jiang, 1983). Moving sand dunes are the product of aeolian sand deposition, and they bury farmland and grassland and destroy vegetation. Being buried by sand affects plant invasion, plant settlement, plant growth, and plant distribution. Strong wind erosion exposes plant seeds buried in the soil, and, thus, influences the process of plant development. Aeolian sand transport destroys plants and vegetation community structure, and even exposes and breaks vegetation roots. Aeolian sand transport affected the processes of photosynthesis and water utilization by plants (Schenk, 1999).

Vegetation can prevent water erosion and wind erosion. Some complex vegetation can result under certain control actions for gravity erosion. Primary vegetation-like moss can play an important role in preventing wind erosion. First, by influencing the wind erosion and sand transport processes, vegetation can control aeolian sand activities. Overground vegetation retains soil moisture, increases the surface roughness, decreases or breaks down the surface wind energy, and absorbs the momentum of saltated sand particles and intercepts sand (van de Ven et al., 1989). Underground root groups of vegetation can fix the soil and improve soil structure (He and Zhao, 2003; Zhang et al., 2009). Second, artificial or natural vegetation fixes sand dunes. With the increase of vegetation coverage, sand dunes are stabilized to semi-fixed and fixed sand deposits. Under the action of nature and biological life, sand deposits experience the soil forming process, without any aeolian sand activities on the surface of sand deposits (Zhang et al., 2009).

Based on field investigations and observations, the structure of aeolian sand flow, wind speed needed to initiate sand particle movement, sediment transport rate, and the vertical distribution of the aeolian sand flow under the action of different vegetation characteristics were quantitatively determined (Zhang et al., 2004; Huang et al., 2001; Liu, 1997). Undisturbed soil samples with natural vegetation and artificial vegetation were examined in wind tunnel experiments and the critical speed for wind erosion, wind velocity profile near the surface, and wind velocity flow field were studied (Zhang et al., 2003a; Zhou et al., 2002; Hu et al., 2002), and the impact of vegetation on the vertical distribution of the sediment transport rate was determined by Dong et al. (1996) and Liu and Dong (2002).

After the sand dunes were stabilized and experienced the soil forming process, the physical, chemical, and biological characteristics change. Evolution characteristics of aeolian sand deposits in desertification areas in northeast China has been studied quantitatively, including grain composition, organic matter content, and microbial content (Liu, 1962; Gu et al., 1999; Xiao et al., 2003).

To quantitatively describe whether the sand dunes were stabilized, scholars have developed different descriptive measures for the degree of dune fixation on the basis of different research emphases. The ratio of vegetation coverage or crusting layer coverage to sand dune area (Shen, 1996; Cui, 1998) and the ratio of sand transport rate to vegetation coverage (Durán and Herrmanm, 2006) have usually been used to describe the degree of dune fixation. In addition, some scholars analyzed measured data of characteristic variables of sand deposits and vegetation, and by using the factor analysis method determined a comprehensive index for judging the degree of dune fixation (Han and Zhang, 2001).

In addition to the effect of vegetation, climate conditions and human activities are also main factors influencing sand dune fixation (Wu, 1987). After the sand dunes are fixed, pioneer plants develop and go through vegetation succession under the influence of biological, environmental, and plant interactions. At the same time, due to different human activities and climate conditions, there is a big difference in aeolian sand deposit characteristics and vegetation structure in different places. Due to different geographical climate conditions and human disturbances, vegetation on sand deposits show different vegetative succession processes (Zhao and Wang, 1999; Chen, 2001; Hrsak, 2004).

Movement of sand dunes and development of vegetation are the main earth surface processes in desertification areas. Moving sand dunes damage vegetation and the ecological environment, and vegetation promotes stabilization of moving sand dunes and ecological restoration of desertification areas. The competition between vegetation and aeolian sand erosion affects and determines the geomorphic tendency of desertification areas. Models, based on the interaction between the development of vegetation cover and aeolian sand erosion, can be used to study the evolution trend of vegetation development and the movement of sand dunes in desertification areas for desert management and ecological restoration.

5.3.2 Vegetation-aeolian erosion model

Since domestic and foreign studies on wind erosion models began in the 1940s and 1950s, many scholars have proposed different forms of wind erosion prediction models, which have been widely used in wind erosion prediction and evaluation. At present, wind erosion models can be divided into three types: empirical models, physical models, and mathematical models. Empirical models are mainly empirical formulas developed based on statistical analyses of experimental simulation results and field observation results, but lack rigorous physical and mathematical bases. Physical models are built focusing on key variables according to the physical mechanisms of the variables in the process of wind erosion. But because the wind erosion process is complex, a lot of physical mechanisms are not yet clear, so it is difficult to objectively reflect the process of wind erosion with physical models. Mathematical models generally are obtained by solving equations for aeolian sand flow dynamics. The equations are very complex and must be simplified step by step to solve. In addition, many parameters of the mathematical models have no clear physical meaning and cannot be determined in practical application (Wu, 2003).

The main typical wind erosion models are the Wind Erosion Equation (WEQ) (Woodruff and Siddoway, 1965), Pasak Wind Erosion Equations, Bocharov model, Texas Erosion Analysis Model (TEAM) (Gregory et al., 1988), Wind Erosion Evaluation Model (WEAM), Revised Wind Erosion Equation (RWEQ), Wind Erosion Prediction System (WEPS) (Hagen, 1991), and others. WEQ and the Bocharov model are empirical models developed from experiments and field observations. TEAM combines a theoretical model and an empirical model to form a simplified process model, but TEAM is not able to fully reflect the wind erosion process. The WEPS system comprehensively sums up the previous research results, yielding a good wind erosion prediction models, including in the main program erosion, weather, crop growth, decomposition, soil, hydrology, and farming subroutines, but still has a number of limitations as pointed out by Yang et al. (2003) and Liao et al. (2004).

Wind erosion prediction research started late in China. Based on wind tunnel simulation experiments and field survey data, some scholars have studied the relation between wind erosion and many factors including wind speed, air relative humidity, soil particle size, soil hardness, vegetation coverage, structural breakage of soil structure, and surface slope. An empirical estimation model of the soil wind erosion amount in a small watershed was established. This model refers to spatial and temporal variable functions of variations in the wind erosion process. It is difficult to apply this model in practice (Dong, 1998).

The processes of wind erosion, sand transport, and sand deposition under the action of the vegetation have been simulated. There are many models for calculating the transport rate under the action of vegetation. Some sand transport models consider the impact of vegetation coverage on the wind speed and wind speed for initiation of sand particle movement (Buckley, 1987; Wasson and Nanminga, 1986; Shi, 2005). Additionally, some scholars studied the wind speed for initiation of sand particle movement considering the impact of a single plant, and simulated wind erosion and sediment movement under the action of single shrubs based on the sediment transport equation (Leenders et al., 2011). In addition, the formation and development processes of sand dunes under the action of vegetation have been simulated by many equations. Iterative coupling calculations of wind shear stress, sediment rate equations, conservation of mass, and a vegetation growth model are frequently used (Luna et al., 2011; Durán et al., 2008).

The growth and distribution of vegetation are influenced by factors such as temperature, light, moisture, and soil nutrients. Vegetation exchanges material, energy, and momentum with the surrounding environment in the process of evolution. There are many plant growth models (Gates, 1980; McMartrie and Wolf, 1983; Zhang and Zhang, 2000; Guo and Yuan, 2000). These models may consider a single variable environmental factor or various complicated factors (Walker et al., 1981; Olson et al., 1985), and considering individual plant or vegetation communities, the simulation models may perform well (Sharpe et al., 1985; Li et al., 2003; Zhang and Yang, 2006). A dynamic vegetation model has stimulated the response of vegetation to environmental factors such as carbon dioxide (CO_2) concentration or climate change (Wang, 2006c). Since the late 1990s, the Dynamic Global Vegetation Model (DGVM) has become the focus among vegetation models, and it can be used to evaluate the effects of climate change on vegetation growth (Wang et al., 2009).

In addition to climate conditions and soil factors, the vegetation growth process is affected by the disturbance of ecological stresses such as natural disasters or human activities, such that the original vegetation development and succession process are changed (Wang et al., 2003a; Wang et al., 2003b; Wang et al., 2004). According to the different actions, the ecological stresses are divided into lethal stress and damage stress. The lethal stress leads to plant death and vegetation coverage reduction. These stresses include forest fires, deforestation, volcanic eruptions, landslides, debris flow, etc. The damage stress refers to a reduction in vegetation vitality but not death of vegetation. These stresses include plant diseases and insect pests, grazing, cyclones, drought, pollution, etc. (Wang et al., 2005a). Regarding the vegetation growth process under the action of ecological stresses, studies have focused on the simulation of the physiological change process of vegetation under the action of single or multiple damage stresses, such as a tree death model, the forest vegetation simulation system (FVS) fire extension model, FVS pest model, etc. (Pedersen, 1998).

Even among the previous models described above, the study of a model based on the interaction between the development of vegetation cover and aeolian sand erosion was rare. A vegetation and aeolian sand coupling model with many parameters, based on the interaction between the vegetation cover and aeolian sand erosion, was studied and established by Li et al. (2009). Parameters represent impacts of water, temperature, soil, and wind, but the key impact factors including ecological stresses and human activities were not qualitatively represented.

Vegetation-erosion dynamics, which studies the laws of evolution of vegetation under the action of various ecological stresses, is a new interdisciplinary science. The model of vegetation-erosion dynamics is based on the interaction between vegetation development and soil erosion, and qualitatively defines the impact of ecological stresses and human activities. This model can be used to evaluate vegetation development, water erosion, and the effects of soil and water conservation measures. The vegetation-erosion dynamics model and the vegetation-erosion chart were developed to predict the tendencies of vegetation and erosion under the impacts of natural stresses and human activities. The developed model and the vegetation-erosion chart have been applied to many watersheds. The proposed governing strategies of the vegetation-erosion dynamics model provide theoretical support for watershed management (Wang et al., 2005b; 2008; Wang, 2006a).

Different from water erosion, aeolian sand erosion and vegetation development contend with each other in the fringes of desertification areas. Moving sand dunes damage the vegetation cover, whereas vegetation promotes stabilization of moving sand dunes and ecological restoration of desertification area. The interaction between vegetation development and aeolian sand erosion follows a law of dynamics and affects the geomorphic tendency of desertification areas. Finally, the result of the interaction between vegetation and erosion determines whether the desert will expand or retreat. Therefore, taking the fringe area of a desert as the research area, the vegetation-erosion dynamics model was applied to the desertification area on the Plateau and a coupling equation of vegetation coverage and aeolian sand erosion dynamics was developed. In this model, vegetation coverage and the amount of aeolian sand erosion are used to represent vegetation development and aeolian sand erosion, respectively. This model can be used to simulate and predict the tendencies of vegetation development and aeolian sand dune movement on the fringes of desertification areas. Then management strategies for the desertification area can be proposed according to the vegetation-aeolian sand erosion chart.

On the fringes of desertification areas, the dynamic process between vegetation development and aeolian sand erosion is influenced by natural stresses and human activities including tree planting, tree cutting, and erosion reduction measures. Natural stresses and human activities play a key role in the evolution process of vegetation and aeolian sand erosion. Based on the vegetation-erosion dynamics model, assuming that the action among the stresses are independent, the coupled differential equations for the vegetation-aeolian sand erosion processes under the action of stresses are obtained as follows:

$$
\begin{cases}
\dfrac{dV}{dt} - aV + cE = V_R + V_\tau \\[2mm]
\dfrac{dE}{dt} - bE + fV = E_\tau + E_S
\end{cases}
\tag{5.1}
$$

in which V represents the vegetation cover, E represents the rate of aeolian sand erosion with the dimension of mass $\cdot$ area$^{-1} \cdot$ time^{-1}, V_R represents positive human stresses (e.g., reforestation) with the dimension of time^{-1}, V_τ represents negative human stresses (e.g., deforestation) with the dimension of time^{-1}, E_τ represents the reduction of aeolian sand erosion by the application of straw checkerboard barriers with the dimension of mass $\cdot$ area$^{-1} \cdot$ time^{-2}, E_S represents the reduction of aeolian sand erosion by the application of sand-fixation measures including sandy gravel cover and sand-protecting barriers with the dimension of mass $\cdot$ area$^{-1} \cdot$ time^{-2}.

The physical meanings of the parameters in the equation are: (1) Parameter a: the increase of vegetation coverage under the action of vegetation with the dimension of time^{-1}. Vegetation retains moisture and nutrients in soil and promotes the weathering process of fine sand, resulting in increases in the coverage and density of vegetation. (2) Parameter c: the reduction of vegetation coverage under the impact of aeolian sand erosion with the dimension of length$^2 \cdot$ mass^{-1}. Aeolian sand erosion damages soil structure and the vegetation roots. Thus, movement of sand dunes destroys vegetation. (3) Parameter b: the increase of the aeolian sand erosion rate under the influence of aeolian sand erosion with the dimension of time^{-1}. Aeolian sand erosion destroys the granular structure in the surface soil of a sand dune and exposes the vegetation roots, resulting in increases in the rate of aeolian sand erosion. In addition, aeolian sand erosion destroys the vegetation and releases the underlying fine sand, which further increases aeolian sand erosion. (3) Parameter f: the decrease of the aeolian sand erosion rate under the action of vegetation with the dimension of mass $\cdot$ length$^{-2} \cdot$ time^{-2}. Vegetation development promotes the stabilization of sand dunes and the soil forming process. The resulting crusting and humus layers protect the sand dunes from aeolian sand erosion.

The theoretical solution for the nonhomogeneous linear ordinary differential Eq. (5.1) is as follows:

$$V = c_1 e^{m_1 t} + c_2 e^{m_2 t}$$
$$+ e^{m_1 t} \int \left[e^{-m_1 t} e^{m_2 t} \int e^{-m_2 t} \left(\frac{d(V_\tau + V_R)}{dt} - b(V_\tau + V_R) - c(E_\tau + E_S) \right) dt \right] dt$$

(5.2)

$$E = c_1 \frac{a - m_1}{c} e^{m_1 t} + c_2 \frac{a - m_2}{c} e^{m_2 t}$$
$$+ e^{m_1 t} \int \left[e^{-m_1 t} e^{m_2 t} \int e^{-m_2 t} \left(\frac{d(E_\tau + E_S)}{dt} - a(E_\tau + E_S) - f(V_\tau + V_R) \right) dt \right] dt$$

(5.3)

In which, c_1 and c_2 are the integral constants determined by the boundary and initial conditions. Indices m_1 and m_2 are given as:

$$m_{1,2} = \frac{1}{2} \left[(a + b) \mp \sqrt{(a + b)^2 - 4(ab - cf)} \right]$$

(5.4)

The parameters a, c, b, and f are important in the vegetation-aeolian sand erosion dynamics and are the basis of the vegetation-aeolian sand erosion chart. They are closely related to climate, soil characteristics, and geomorphic conditions, and are not related to vegetation and the erosion rate. That is, the parameters of the vegetation-aeolian sand erosion dynamics model and the vegetation-erosion chart are the same in deserts with the same climate and landform conditions. Based on the vegetation-aeolian sand erosion dynamics model, the parameters could be obtained by a trial-and-error method and measured data. First, the vegetation coverage, V, aeolian sand erosion rate, E, vegetation ecological stress, V_τ and V_R, and reduction of the aeolian sand erosion rate, E_τ and E_S, of many years are calculated from measured data. Second, the coupled differential equations of vegetation-aeolian sand erosion dynamics [Eq. (5.1)] are adapted to the difference equations with the differential unit of one year. Third, the parameters a, c, b, and f could be obtained by a trial-and-error method and measured data. Finally, the evolution of vegetation and aeolian sand erosion in the research area could be studied by the vegetation-aeolian sand erosion chart. According to the results, suggestions could be provided for the management and ecological restoration in the desertification area. The following subsection presents the application of the vegetation-aeolian sand erosion dynamics model.

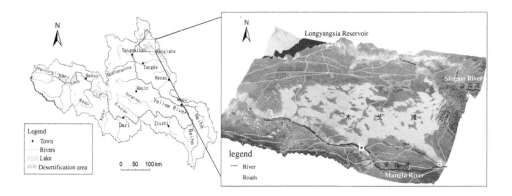

Figure 5.10 Mugetan desertification area in the source area of the Yellow River.

5.3.3 Application of the model to the Mugetan desert

5.3.3.1 The Mugetan desertification area

The Mugetan desertification area is located in the Gonghe basin in the source area of the Yellow River with an area of 790 km². Figure 5.10 shows the Mugetan desertification area and surrounding streams. Widely deposited sickly yellow or pale brown silt-sand terrane is present in the Gonghe Basin and the sediment deposition was confirmed to be about 1500 m thick by geophysical exploration (Xu and Xu, 1983). The main planation surface of the Plateau thoroughly collapsed in the uneven uplift due to the movement of the Plateau, and the vertical deformation was up to 1700 m. The Gonghe movement caused the Yellow River to enter the Gonghe Basin 0.11 million years ago, and then the Yellow River eroded the basin at an average rate of 3.5 mm per year. At the same time ancient alluvial and diluvial fans at the edge of the basin rose at a similar rate. At last a layered landform system developed with about 2000 m of elevation difference.

The Yellow River and its tributaries have incised the Plateau for a long time, and multistage terraces with elevations from 3000 to 2200 m above sea level have formed. The Mugetan desertification area is located in a high level terrace without any influence by water erosion. The Mugetan desertification area has a typical plateau continental climate. The average temperature from 1961 to 2010 is 2.4°C, the average sunlight exposure from 1961 to 2010 is 2,720 hours, average rainfall over from 1961 to 2010 is 400 mm, and the average evaporation over from 1961 to 2010 is 1500 mm. The prevailing wind direction is southeast. The maximum wind speed over from 1961 to 2010 is 14 m/s per year, and the annual average number of high wind (≥ 7 m/s) days is 12 days (Guo et al., 2009; Guo et al., 2010).

Figure 5.11 shows the sequential values of the number of high wind days and the maximum high wind values for each year from 1961 to 2009. In the fringe area of the Mugetan desertification area, aeolian sand disasters are severe and have become a serious threat to ecology and pasture land. Vegetation restoration projects have been applied in the fringe area of the Mugetan desertification area since the 1970s, mainly including afforestation, laying grass squares, and sandy gravel sand-fixation. Since

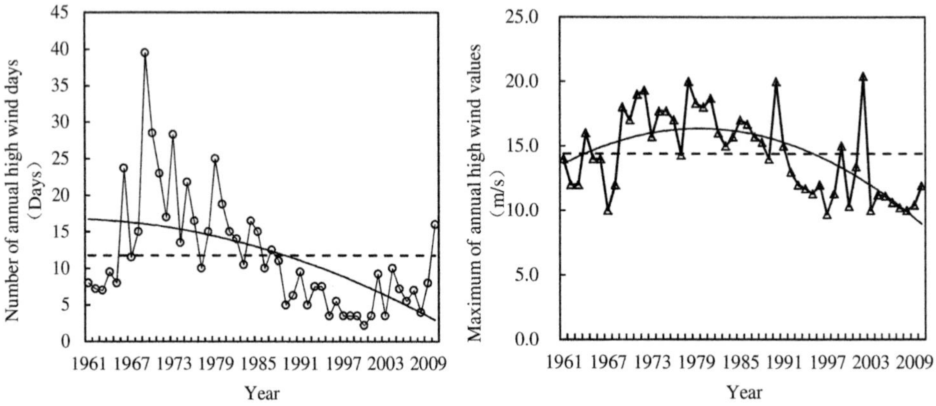

(a) Number of high wind days

(b) Maximum wind velocity

Figure 5.11 Windy days and wind velocity in the period from 1961 to 2009.

then vegetation growth and aeolian sand erosion contended with each other in the fringe area of the Mugetan desertification area. The desertification mitigation effect depends on whether vegetation coverage can control wind erosion in the fringe area of the Mugetan desertification area. Once the movement of sand dunes is controlled by vegetation within a certain region, the desertification area will shrink, and finally the ecological environment of the desertification area will recover and improve to a good condition. Moving sand dunes, semi-fixed sand dunes, and fixed sand deposits are widely distributed in the fringe area of the Mugetan desertification area. The vegetation coverage of moving sand dunes is less than 5% (Fig. 5.12(a)). The surface sand of a moving dune is easily moved by wind. The vegetation cover of a fixed sand deposit is over 70–80% (Fig. 5.12(b)). A humus layer with a depth of 1 to 3 cm on a fixed sand deposit can effectively stop aeolian sand erosion. That is, the surface sand does not move. The vegetation coverage of semi-fixed sand dunes is 5%–60%. Semi-fixed sand dunes with a vegetation coverage of 5%–30% are mainly covered with trees and shrubs (Fig. 5.13(a)), and the surface sand is easily moved. Semi-fixed sand dunes with a vegetation coverage of 30%–60% are mainly covered with trees, shrubs, straw checkerboard barriers, and natural restored vegetation (Fig. 5.13(b)).

Based on the remote sensing images of more than 20 years, including a MultiSpectral Scanner (MSS) image (1977), Thematic Mapper (TM) images (1988, 1996, 2003, 2005, 2007, 2008, 2009, 2010), a SPOT 2/4 image (2006), and 43 field ground-object identification spots (widely distributed around the Mugetan desertification area), the ground features of ten years were obtained. The area values of the ground features were determined by the Environment for Visualizing Images (ENVI) and an ArcGIS system (Ma et al., 2011; Han et al., 2009). Ground features in the fringe area of the Mugetan desertification area include moving sand dunes, semi-fixed sand dunes with a vegetation coverage of 5%–30%, semi-fixed sand dunes with a vegetation coverage of 30%–60%, fixed sand deposits with high vegetation coverage, meadows, areas of

Figure 5.12 Mugetan desertification area (a) Moving sand dunes; and (b) Fixed sand dunes.

(a) Vegetation coverage of 5%~30%   (b) Vegetation coverage of 30%~60%

Figure 5.13 Semi-fixed sand dunes in the Mugetan desertification area.

bare soil, and open water areas. The regions of fixed sand deposits with high vegetation coverage have coverage of 70%–80% without any wind erosion. The region of bare soil with developed vegetation roots has a much lower aeolian sand erosion rate than the region of aeolian sand dunes. Therefore, the aeolian sand erosion rate in the region of bare soil was ignored in the calculation of the aeolian sand erosion rate of the Mugetan desertification area. The range of the fringe area of the Mugetan desertification area was identified based on the ground features measured between 1977 and 2010 with an area of 423.9 km^2. Figures 5.14(a) and 5.14(b) show sketch maps of site a and site b in Fig. 5.10. Figure 5.15 shows feature photos on the edges of the study area.

5.3.3.2 Determination of parameters of the vegetation-erosion dynamics model

In the equations of the model of vegetation-aeolian sand erosion dynamics, the aeolian sand erosion rate, E, is the wind erosion sediment load on an unit area per year,

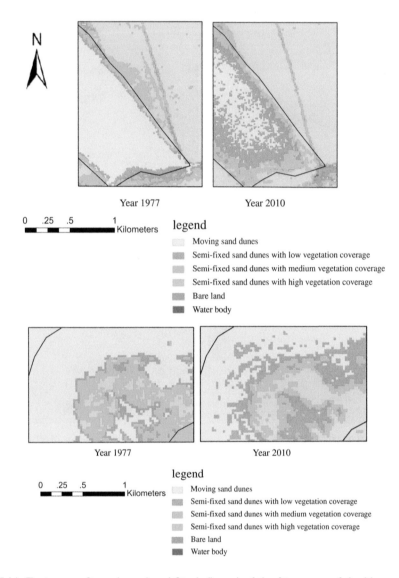

Figure 5.14 Zoning map Site a (upper) and Site b (lower) of the fringe area of the Mugetan Desert (location of a and b are showing in Fig. 5.10).

regardless of the sand transport distance and accumulation process. Vegetation coverage, V, is the vegetation-covered area on an unit area, and can be used to evaluate the development state of vegetation in the study area. Records of aeolian sand erosion and vegetation in the Mugetan desertification area were rare. The values of vegetation coverage over many years were approximately calculated using the remote sensing images previously listed. Furthermore, the aeolian sand erosion rates over many years were approximately calculated using remote sensing images and measured aeolian sand

(a) Junction of desert and grassland   (b) Junction of the desert core area and the fringe area

Figure 5.15 Fringe area of the Mugetan desertification area.

erosion depths. From 1977 to 2010 in the fringe area of the Mugetan desertification area, aeolian sand erosion and vegetation development fiercely contended with each other. The dynamic process between vegetation development and aeolian sand erosion is influenced by natural stresses and human activities including tree planting, tree cutting, and erosion reduction measures including grass squares, sandy gravel, and sand-fixation grass squares. Afforestation, which is a positive ecological stress, prevents wind erosion, fixes sand dunes, and increases vegetation coverage. Considering engineering documents of sand-fixation projects and field investigation of afforestation areas, planting density and a single estimation of vegetation coverage were estimated. Deforestation, which is a negative ecological stress, reduces the vegetation coverage. Considering previous documents and field investigations, the area and the extent of vegetation coverage reduction were estimated. Grass squares, gravel and sandy measures, and sand-protecting barriers resist the movement of sand dunes and reduce the amount of wind erosion. By using engineering data and documents, the wind erosion depth and the reduction of the aeolian sand erosion amount were estimated. Forest reservations reduce the outside impact on vegetation by building isolation areas, providing a benefit to the development of vegetation. These vegetation processes are reflected in the parameter *a* of the equations of the model of vegetation-aeolian sand erosion dynamics.

Based on the revised remote sensing data of many years, vegetation coverage and the area values of different locations were calculated using ArcGIS and ENVI software (Ma et al., 2011; Han et al., 2009; Li et al., 2009). Aeolian sand erosion depths were obtained by measuring the erosion depth around the vegetation roots. In the region of moving sand dunes, aeolian sand erosion depths of 4 locations on the surfaces of 3 moving sand dunes were measured. The annual erosion rate of aeolian sand was roughly 15 cm/yr. In the region of semi-fixed sand dunes with vegetation cover of trees and shrubs, aeolian sand erosion depths of 14 locations on the surfaces of 6 semi-fixed sand dunes were measured. The aeolian sand erosion depths of 5 years were

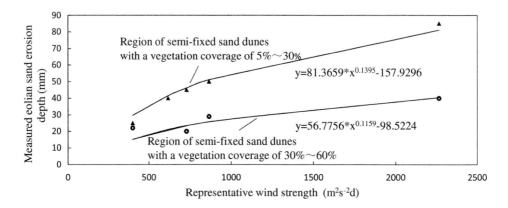

Figure 5.16 Annual aeolian sand erosion depth as a function of the representative wind strength.

obtained and used to estimate the aeolian sand erosion depths of the region of semi-fixed sand dunes with a vegetation coverage of 5%–30%. In the region of semi-fixed sand dunes with vegetation cover of trees, shrubs, and straw checkerboard barriers, aeolian sand erosion depths of 11 locations on the surfaces of 4 semi-fixed sand dunes were measured. The aeolian sand erosion depths of 4 years were obtained, and used to estimate the aeolian sand erosion depths of the region of semi-fixed sand dunes with a vegetation coverage of 30%–60%. Based on the meteorological and landform data, the annual aeolian sand erosion depths from 1977 to 2010 were calculated by a curve fitting calculation. The fringe area of the Mugetan desertification area is far away from agricultural production areas and areas where people are living and, thus, is not interfered by human activities.

The aeolian sand erosion depths in the fringe areas of the Mugetan desertification area are approximately related with climate conditions. Wind tunnel tests and field measurements show that the greater the wind speed and number of high wind days are, the greater the wind erosion depth is (Yao et al., 2001). The wind erosion amount has a quadratic relation with wind speed, a negative quadratic power function relation with soil particle size, and an exponential function relation with vegetation coverage. Figure 5.16 shows the fitted curves of aeolian sand erosion depths changes as a function of the representative wind strength since 2006. The abscissa is the representative wind strength which is equal to the product of square of the annual maximum wind speed and annual number of high wind days. Based on the meteorological data at Guinan and the fitting formula given in Fig. 5.16, the wind erosion depth values since 1961 were obtained. Based on the wind erosion depth values and the area values of different locations, the annual aeolian sand erosion rate was roughly calculated as listed in Table 5.3.

Beginning in the late 1970s, forest planting and afforestation were performed through the Three North Shelterbelt Program (China Forest Construction Bureau, 2009). Beginning in the late 1990s, many control measures, such as forest planting, afforestation, and straw checkerboard barriers were applied through several plans (Yang et al., 2006; Ma, 2006, Wang et al., 2000). From 1998 to 1999, people cut down

Table 5.3 Estimated values of the aeolian sand erosion rate and vegetation coverage in the fringe area of the Mugetan desertification area.

Year	1977	1988	1996	2003	2005	2006	2007	2008	2009	2010
Aeolian sand erosion rate (E) (t · km^{-2} · yr^{-1})	100876	91513	77562	66874	56037	52747	50814	47975	48201	38214
Vegetation coverage (V) (%)	7.00	12.87	16.77	23.47	24.38	25.76	31.30	31.87	32.59	35.88

Table 5.4 Annual value of ecological stress and reduction of the aeolian sand erosion rate.

Year	Vegetation stress V_τ, V_R (time^{-1})	Reduction of aeolian sand erosion rate E_τ, E_S (mass · area^{-1} · time^{-2})
1977–1996	$V_R = 0.2$–0.5%	$E_\tau + E_S = 4000$–6000 t · km^{-2} · a^{-1}
1997–1999	$V_R = 0.2$–0.5%; $V_\tau = 0.5$–1.0%	$E_\tau + E_S = 4000$–6000 t · km^{-2} · a^{-1}
2000–2005	$V_R = 0.2$–0.5%	$E_\tau + E_S = 4000$–7000 t · km^{-2} · a^{-1}
2006–2010	$V_R = 0.5$–1.0%	$E_\tau + E_S = 5000$–8000 t · km^{-2} · a^{-1}

and dug down seriously without plans (La et al., 2001; La, 2002). In the 2000s, a series of management projects was continuously performed. Measures including planting trees, straw checkerboard barriers, sandy gravel cover, sand-protection barriers, and forest reservation measures were combined to stabilize the sand dunes and restore the ecological environment. The annual value of vegetation ecological stress, V_τ and V_R, and the reduction of the aeolian sand erosion rate (E_τ and E_S) were approximately estimated based on project data, literature values, and measured data. Table 5.4 lists the results.

Based on the previously calculated results including vegetation coverage, aeolian sand erosion rate, value of vegetation ecological stress, and reduction of the aeolian sand erosion rate, a trial-and-error method was performed until the best-fitting values of the parameters of the vegetation-erosion dynamics model were obtained. The parameters for the Mugetan desertification area were determined as follows:

$$a = 0.06; \quad c = 0.0000000987; \quad b = 0.125; \quad f = 16000 \qquad (5.5)$$

For a comparison of the model and measurements, Fig. 5.17 shows the measured and calculated processes of vegetation development and aeolian sand erosion of the Mugetan desertification area.

5.3.3.3 Applications of the vegetation-erosion chart in the Mugetan desert

According to the model of the vegetation-aeolian sand erosion dynamics (Wang et al., 2003b) the vegetation-aeolian sand erosion chart was obtained. In the case of no human-induced stresses, assume that the stress terms in Eq. (5.1) are equal to zero and $V' = \frac{dV}{dt}$, $E' = \frac{dE}{dt}$ can be rewritten as $V' = 0$, and $E' = 0$. V' and E' may be positive or

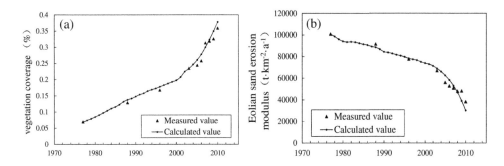

Figure 5.17 Comparison of the measured and calculated processes of (a) vegetation coverage and (b) aeolian sand erosion rate in the fringe area of the Mugetan desertification area.

negative, therefore, the V-E plane: $V \in [0,1]$, $E \in [0, \infty)$ can be divided into 3 zones by the two lines $V' = 0$, $E' = 0$. The three zones are: Zone A: $dV/dt < 0$, $dE/dt > 0$, in this zone the vegetation cover is deteriorating and the erosion rate is increasing. Zone C: $dV/dt > 0$, $dE/dt < 0$, in this zone the vegetation cover is increasing and the erosion rate is deteriorating. Zone B: $dV/dt > 0$, $dE/dt > 0$ or $dV/dt < 0$, $dE/dt < 0$. The vegetation-erosion chart is used to discuss the development trend of vegetation and erosion in the case of no human-induced stresses. From equation (5.1), two lines $V' = 0, E' = 0$ which divide the V-E plane into three parts depend on the four parameters a, c, b, and f as follows:

$$E = \frac{a}{c}V; \quad E = \frac{f}{b}V \tag{5.6}$$

Figure 5.18 shows the vegetation-aeolian sand erosion chart for the Mugetan desertification area. The curve is the evolution process of the vegetation coverage and aeolian sand erosion rates over twenty three years.

The vegetation-aaeolian sand erosion chart of the source region of the Yellow River has a relatively large C-Zone. Figure 5.18 indicates that once the vegetation coverage reaches a certain value and the aeolian sand erosion rate reduces to a certain value, the vegetation in the source region of the Yellow River will have a strong ability to self-improve, whether through the natural vegetation development process or through an artificial vegetation succession process. As long as the vegetation does not suffer severe damage, the vegetation will develop well and the sand dunes will stabilize.

The ecological system of vegetation coverage and aeolian sand erosion in the fringe area of the Mugetan desertification area in 1977 is in Zone A. The intensity of aeolian sand erosion is far greater than the protective action of vegetation. In Zone A, the vegetation cover is deteriorating and the erosion rate is increasing. Then, the desertification is getting worse. After 1977, trees were planted and straw checkerboard barriers were installed. When the vegetation cover increased to 15% and the aeolian sand erosion rate reduced to less than 85,000 $(t \cdot km^{-2} \cdot y^{-1})$, and the vegetation could preliminarily control the moving sand dunes. In Zone B, the vegetation cover is in an unstable state. Both vegetation and erosion are increasing. If erosion increases faster

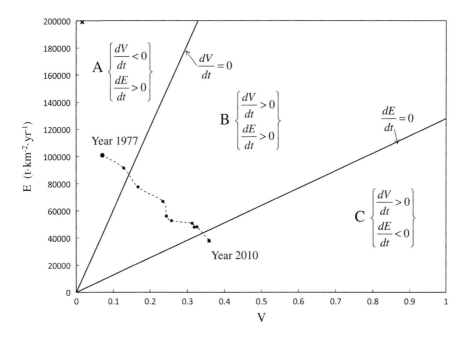

Figure 5.18 Vegetation-erosion chart for the Mugetan desertification area.

or human-stresses cause deforestation and erosion continues to increase the ecological system may enter Zone A. If vegetation increases faster or human controls are applied to erosion, such as reforestation of the hills, the ecological system may enter Zone C. Since the 1980s, the vegetation projects have become effective in the fringe area of the Mugetan desertification area. The vegetation coverage has increased year by year, and the aeolian sand movement has been gradually restrained. When the vegetation cover increases to 35% and the aeolian sand erosion rate reduces to less than $40,000\,\mathrm{t \cdot km^2 \cdot yr^{-1}}$, the ecological system enters Zone C. In Zone C, the ecological system is developing toward complete vegetation cover and an aeolian sand erosion rate of zero. Figure 5.18 shows that the ecological system presently is on the edge of Zone C assuming no change in status since 2010. Sand dunes are under control by vegetation, but are not stable. Vegetation management should still be intensively performed until the vegetation develops to rapidly stabilize the aeolian sand dunes. Then under the action of developed vegetation, the sand dunes in the fringe area of the Mugetan desertification area will gradually be stabilized and the expansion of desert will be stopped.

The state of point "x" which is a long way from Zone C in Fig. 5.18 represents most places of the Mugetan desertification area. The "x" state cannot enter Zone C by afforestation. At present, it is necessary to control aeolian sand erosion and push the point "x" state to move down and to the right by many measures. When the aeolian sand erosion rate is reduced to less than $100,000–85,000\,\mathrm{t \cdot km^2 \cdot y^{-1}}$, vegetation management could make the ecological systems of most places of the Mugetan desertification area enter Zone C. When the aeolian sand erosion rate is reduced to less than

$40,000 \, t \cdot km^2 \cdot y^{-1}$, vegetation measures should be the main control measures. The vegetation-aeolian sand erosion chart also shows that, desert management could start from the edges of the desert and gradually advance toward the center of the desert. First the ecological system of the fringe area of desert must enter Zone C, and then the management strategies may further advance toward the desert core step by step.

REFERENCES

Abolkhair Y. M. S. 1986. The statistical analysis of the sand grain size distribution of Al-Ubaylah barchan dunes, northwestern Ar-Rub-Alkhali Desert, Saudi Arabia. GeoJournal, 13(2), 103–109.

Bendali F., Floret C., Floch E.L., Pantanier R. 1990. The dynamics of vegetation and sand mobility in arid regions. Journal of Arid Environments, 18, 21–32.

Buckley R. 1987. The effect of sparse vegetation on the transport of dune sand by wind. Nature, 325, 426–428.

Buckley R. 1996, Effects of vegetation on the transport of dune sand. Annals of Arid Zone, 35(3), 215–223.

Chen Longheng. 1998. Aeolian Sandy Soil in China. Beijing: Science Press, p. 32–33. (In Chinese).

Chen Shengyong. 2001. Summarizing study achievement of vegetation succession on the sand land. Soil and Water Conserbation Science and Technology in Shanxi, (4), 23–26. (In Chinese).

Cui Xun. 1998. The evolution of sand dunes and its main factor change. Arid Zone Research, (2), 47–49. (In Chinese).

Dai Junhu, Ge Quansheng, Wang Mengmai. 2010. Fragile ecosystem restoration and regional sustainable development in the Sanjiangyuan region – a brief introduction of most project. In: Landscape and Environment Science and Management in the Sanjiangyuan Region. Gary Brierley, Li Xilai, Chen Gang, eds. Qinghai: Qinghai People's Publishing House, p. 221–227.

Dong Guangrong, Li Changzhi, Jin Jiong, and Gao Shangyu. 1987. The results of wind erosion by wind tunnel simulation. Chinese Science Bulletin, (4), 297–301. (In Chinese).

Dong Suocheng, Zhou Changjin, and Wang Haiying. 2002. Ecological crisis and countermeasures of the Three Rivers' Headstream Regions. Journal of Natural Resources, (6), 713–720. (In Chinese).

Dong Zhibao. 1998. Establishing statistic model of wind erosion on small watershed basis. Bulletin of Soil and Water Conservation, 18(5), 55–62. (In Chinese).

Dong Zhibao, Chen Weinan, Li Zhenshan. 1996. The experimental study of the effect of vegetation on soil wind erosion. Journal of Soil Erosion and Soil and Water Conservation, 2(2), 1–9. (In Chinese).

Durán O., Herrmann, H. J. 2006. Vegetation against dune mobility. Physical Review, 97(18), 1–4.

Durán O, Silva M.V.N., Bezerra L.J.C., Herrmann H.J., Maia L.P. 2008. Measurements and numerical simulations of the degree of activity and vegetation cover on parabolic dunes in north-eastern Brazil. Geomorphology, 102, 460–471.

Fang Xiaomin, Han Yongxiang, Ma Jinhui, Song Lianchun, Yang Shengli, Zhang Xiaoye. 2004. Characteristics of sand and dust in Qinghai-Tibet Plateau and the plateau loess accumulation-A case study: The dust weather process in Lhasa. Chinese Science Bulletin, 49(11), 1084–1090. (In Chinese).

Feng Jianmin, Wang Tao, Xie Changwei. 2004. Land degradation in the source region of the Yellow River: Northeast Qinghai-Xizang Plateau. Ecology and Environmental, 13(4), 601–604. (In Chinese).

China Forest Construction Bureau. 2009. Atlas of Desertification and Desertification Land in China. Beijing: Science Press, p. 62–74. (In Chinese)

Fu Biqian. 2006. Principle and Methods of Ecological Experience. Beijing: Science Press. (In Chinese).

Gates D.J. 1980. Competition between two types of plants located at random on a lattice. Math Biosci, 48(3), 157–194.

Gregory J.M., Borrelli J., Fedler C.B. 1988.TEAM: Texas erosion analysis model. Proceedings, 1988 Wind Erosion Conference. Texas Tech University, Lubbock, Texas, p. 88–103.

Gu Fengxue, Wen Qikai, Pan Borong. 1999. Review on aeolian sand soil development trend research under impact of artificial vegetation. Arid Zone Research, 16(2), 67–70. (In Chinese).

Guo Lianyun, Zhong Cun, Ding Shengxiang, Han Huifu. 2009. Local climate changes and their impacts on grassland degradation in Gonghe basin of Guinan county of Qinghai province in past half-century. Chinese Journal of Agrmoeteorology, 230(2), 147–152. (In Chinese).

Guo Ruihai, Yuan Xiaofeng. 2000. Hopf bifurcation for a ecological mathematical model on microbe populations. Applied Mathematics and Mechanics. 21(7), 693–700. (In Chinese).

Guo Shousheng, He Lianbing., Xu Zhengfu. 2010. Variety trend analysis of sunshine hours in Guinan county in recent 50 Years. Journal of Anhui Agricultural Sciences, 38(16), 8530–8532, 8538. (In Chinese).

Hagen L.J. 1991. A wind erosion prediction system to meet the users need. Journal of Soil and Water Conservation, 46(2), 107–111.

Han Guang, Zhang Guifang. 2001. A quantitative analysis on fixing extent of aaeolian sand-dunes. Acta Ecologica Sinica, 21(7), 1057–1063. (In Chinese).

Han Haihui, Yang Taibao, Wang Yilin. 2009. Dynamic analysis of land use and landscape pattern changes in Guinan county, Qinghai, in the past 30 years. Process in Geography, 28(2), 207–215.

Han Yongxiang, Xi Xiaoxia, Song Lianchun, Ye Yanhua, Li Yaohui. 2004. Spatio-temporal sand-dust distribution in Qianghua-Tibet Plateau and its climatic significance. Journal of China Desert Research, 24(5), 588–592. (In Chinese)

He Binghui, Liu Lizhi. 2007. Study on the cracking process of purple shale and the charaeteristies of alluvial purple soil developed from suining group. Journal of Southwest University (Natural Science), 29(1), 48–52. (In Chinese)

He Zhibing, Zhao Wenzhi. 2003. Characteristics of soil moisture of different vegetation types in initial stage of fixed sand dune of semi-arid region, Journal of Soil and Water Conservation, 17(4), 164–167. (In Chinese).

Houghton J. 1998. Global Warming. Dai Xiaosu, ed. Beijing: Meteorological Press, Beijing, p. 120–127.

Hrsak V. 2004. Vegetation succession and soil gradients on inland sand dunes. Ecology, 23(1), 24–39.

Hu Guangyin, Dong Zhibao, Wei Zhenhai, Man Duoqing, and Wang Wenli, 2008. Progresses and perspective of studies on sandy desertification in the source regions of Yangtze River, Yellow River and Lantsang River. Journal of Arid Land Resources and Environment, 22(3), 41–44.

Hu Mengchun, Zhao Aiguo, Li Nong. 2002. Sand-trapping efficiency of railway protective system in Sapotou tested by wind tunnel. Deserts of China, 22(6), 598–601. (In Chinese).

Huang Fuxiang, Niu Haishan, Wang Mingxing, Wang Yuesi, Ding Guodong. 2001. The relationship between vegetation cover and sand transport flux at Mu Us sandland. Acta Geographica Sinica, 56(6), 700–710. (In Chinese).

Jiang Jin. 1983. Research on biology and physiology characteristics of main sand-fixation plant in Shapotou, Forest Science, 19(2), 113–120. (In Chinese)

Kang Zhicheng, Li Zhoufen, Ma Ainai. 2004. Research of debris flows in China. Beijing: Science Press, p. 79–81.

Khatelli H., Gabriels D., Wang Xiu. 1998. Dynamic study of Tunisia dunes–Barchans from Tunisia dunes move towards the Sahara Desert. Soil and Water Conservation Science and Technology Information, 3, 20–23.

La Yuanlin. 2002. The current situation of grassland ecological environment and its control strategies. Pratacultural Science, 19(6), 1–4. (In Chinese).

La Yuanlin, Liang Zhuying, Banma Duojie. 2001. Alpine grassland desertification and control method and countermeasure of Guinan county. Qinghai Prataculture, 11(1), 41–45. (In Chinese).

Leenders J. K., Sterk G., Van Boxel J. H. 2011. Modelling wind-blown sediment transport around single vegetation elements. Earth Surface Processes and Landforms, (36), 1218–1229.

Li Yanfu, Wang Zhaoyin, Shi Wenjing, Wang Xuzhao. 2010. Slope debris flows in the Wenchuan Earthquake area. Journal of Mountain Science, (7), 226–233.

Li Zizhen, Wang Wanxiong, Xu Cailin. 2003. The dynamic model of crop growth system and numerical simulation of crop growth process under the multi-environment external force action. Applied Mathematics and Mechanics, 24(6), 644–652. (In Chinese).

Li Zhenshan, Wang Yi, He Limin. 2009. Vegetation-erosion process in semiarid region: I. Dynamical Models, Journal of Desert Research, 29(1), 23–30. (In Chinese)

Liao Ciyuan. 1980. Research on the choice machine characteristics of several sand-fixation plants, In: Quicksand Quality Research, Ningxia: Ningxia People's Publishing House, p. 60–120. (In Chinese).

Liao Chaoying, Li Jing, Zheng Fenli, Liu Guobin. 2004. Research history and trend of wind erosion prediction abroad. Research of Soil and Water Conservation, 11(4), 50–53. (In Chinese).

Lin Chaofeng, Chen Zhanquan, Xue Quanhong, Lai Hangxian, Chen Laisheng, Zhang Dengshan. 2007. Nutrient contents and microbial populations of Aaeolian sandy soil in Sanjiangyuan region of Qinghai Province. Chinese Journal of pplied Ecology. 28(1), 101–106. (In Chinese).

Liu Qianzhi. 1997. The impact of vegetation type on the sand flow structure in Jingdian irrigation area. Gansu Forestry Science and Technology, (3), 13–17. (In Chinese).

Liu Xiaoping, Dong Zhibao. 2002. Wind tunnel tests of roughness and drag partition on vegetated surfaces. Journal of Desert Research, 22(1), 82–87. (In Chinese).

Liu Yumin. 1962. The growth the sandy soil in Qaidam basin, Chinese Journal of Soil Science, (4), 45–48. (In Chinese).

Liu Zhimin, Zhao Xiaoying, Liu Xinmin. 2002. The relation between interference and vegetation. Acta Prataculturae Sinica, 11(4), 1–9. (In Chinese)

Liu Hujun, Zhao Ming, Wang Jihe, Xu Xianying, Liao Kongtai, Wei Huaidong. 2005. Geomorphology characteristics of wind-drift sands in south of Kumtag Desert. Journal of Arid Land Resources and Environment, 17, 130–134. (In Chinese)

Luna M.C.M.D.M, Parteli E.J.R, Durán O., Herrmann H.J. 2011 Model for the genesis of coastal dune fields with vegetation. Geomorphology, 129, 215–224.

Ma Jianhua. 2006. An ecological oasis with an area of 18 million mu was completed in Guinan. Qinghai News Network, http://news.sina.com.cn/c/2006-07-06/08449385326s.shtml. (In Chinese).

Ma Yufeng, Yan Ping, Zhang Dengshan. 2011. Pattern of aaeolian-fluvial interaction in Gonghe basin, Qinhai Province based on GIS. Journal of Arid Land Resources and Environment, 25(1), 151–156. (In Chinese).

McMartrie R., Wolf L. 1983. A model of competition between trees and grass for radiation, water and nutrients. Annals of Botany, 52(4), 449–458.

Niu Zhan, Li Jing, He RuiLi, Ji Junfeng. 2006. The correction method of grain size of total load for the phenomena of results of two series of grian size analysis intercorssed during the analysis with both sieve method and laser grain size analyzer. Journal of China Hydrology, 26(1), 72–75. (In Chinese)

Olson R.L.J., Sharpe P.J.H., Wu H. 1985. Whole-plant modelling: A continuous-time Markov (CTM) approach. Ecological Modelling, 29, 171–187.

Pedersen B.S. 1998. Modeling tree mortality in response to short- and long-term environmental stresses. Ecological Modelling, 105, 347–351.

Qi Yanbing, Chang Qingrui, Wei Xin, Zhang Jing. 2005. Effect of artificial vegetation restoration on sandy soil characteristics in high frigid regions of China. Chinese Agricultural Science Bulletin, (8), 404–408. (In Chinese).

Ren Zhenqiu. 1990. Global Change. Beijing: Science Press, p. 193–213. (In Chinese)

Schenk H.J. 1999. Clonal splitting in desert shrubs. Plant Ecology, 141, 41–52.

Sharpe P.J.H, Walker J., Penridge L.K., Wu H. 1985. A physiologically based continuous-time Markov approach to plant growth modelling in semi-arid woodlands. Ecological Modelling, 29, 189–213.

Shen Weishou. 1996. Fixed process of moving sand dunes in Maowusu. Journal of Soil Erosion and Soil and Water Conservation, 2(1), 17–21. (In Chinese).

Shi Hongzhi. 1991. Reports of Global Environment. China Environmental Science Press, p. 55, 87–89, 102–103. (In Chinese).

Shi Xuefeng. 2005. Relation between vegetation conditions and aeolian sand activities in north semiarid regions. Master's Thesis. Beijing: Life and Environmental Science College of Central University for Nationalities. (In Chinese)

State Forestry Bureau (SFB). 2009. The 30th anniversary summary report of the north shelter forest system construction in Qinghai province [N/OL].China Forestry Resources, (2009-6-19)[2011-9-1]. http://www.forestry.gov.cn/portal/sbj/s/2656/content-422070.html.

State Forestry Bureau (SFB). 2005. China desertification and desertification situation communiqué. (In Chinese).

Tao Zhen, Dong Guangrong. 1994. The relationship between land desertification and climate change in the Guinan desertification area since the last glacial age. Journal of Desert Research. (2), 42–48.

United Nations Environment Programme (UNEP). 2006. Global Environment Outlook–Global Deserts Outlook, June, City of Publication.

van de Ven T.A.M, Fryrear D.W, Spaan W.P. 1989. Vegetation characteristics and soil loss by wind. Journal of Soil and Water Conservation, 44, 347–349.

Walker B.H., Ludwig D., Holling C.S., Peterman R.M. 1981. Stability of semi-arid savanna grazing systems. Journal of Ecology, 69, 473–498.

Wang Feixin. 2006a. Research on the vegetation-erosion dynamics and its application to typical erosion areas of China. Doctoral Dissertation, Tsinghua University, China.

Wang Genxu, Li Qi, Chen Guodong, Shen Yongping. 2001. Climate change and its impact on the eco-environment in the source region of the Yangtze and Yellow rivers in recent 40 years. Journal of Glaciology and Geocryology, 23(4), 346–352. (In Chinese).

Wang Qing. 2006b. The United Nations Environment Programme report: Global desertification threat increases. First Financial Daily, (2006b-6-6) [2011-03-27]. http://news.qq.com/a/20060606 /000073.htm. (In Chinese)

Wang Qingchun, Zhou Lusheng. 1998. Diagnostic analysis of climate change in the source area of Yangtze River and Yellow River. Journal of Qinghai Environment, 8(2), 73–77.

Wang Tao. 2004. Progress in sandy desertification research of China. Journal of Geographical Sciences, 14(4), 387–400. (In Chinese).

Wang Xuemei. 2006c. The international development trend of dynamic vegetation model. Science Journalism, (5), 18–20. (In Chinese).

Wang Xufeng, Ma Mingguo, Yao Hui. 2009. The research progress of dynamic global vegetation model. Journal of Remote Sensing Technology and Application, 24(2), 246–251. (In Chinese).

Wang Zhaoyin, Wang Guangqian, Gao Jing. 2003a. An ecoligical dynamics model of vegetation evolution in erosion area. Acta Ecologica Sinica, 23(1), 98–105. (In Chinese).

Wang Zhaoyin, Wang Guangqian, Huang Guohe. 2008. Modeling of state of vegetation and soil erosion over large areas, International Journal of Sediment Research, 23(3), 181–196.

Wang Zhaoyin, Guo Yanbiao, Li Changzhi, Wang Feixin. 2005b. Vegetation-erosion chart and its application in typical watershed in China. Advances in Earth Science, 20(2), 149–157. (In Chinese).

Wang, Zhaoyin, Huang, Guohe, Wang, Guangqian, and Gao, Jing, 2004, Modeling of vegetation-erosion dynamics in watershed systems. Journal of Environmental Engineering, 130(7), 792–800.

Wang Zhaoyin, Wang Guangqian, Li Changzhi, Wang Feixin. 2003b. Preliminary exploration and application of the vegetation and erosion dynamics. Science in China (Series D) Earth Sciences, 33(10), 1013–1023. (In Chinese).

Wang Zhaoyin, Wang Guangqian, Li Changzhi, Wang Feixin. 2005a. A preliminary study on vegetation-erosion dynamics and its applications. Science in China (Series D) Earth Sciences, 48(5), 689–700. (In Chinese).

Wang Zhitao, Zhu Chunyun, Yang Zhanwu, Chen Wusheng. 2000. A comprehensive summary of control techniques and experience of moving sand dunes in Huangshatou in Guinan. Science and Technology of Qinghai Agriculture and Forestry, (3), 45–47. (In Chinese).

Wasson R.J., Nanninga P.M. 1986. Estimating wind transport sand on vegetated surface. Earth Surface Processes and Landforms, 11(4), 505–514.

Wolfe S.A., Nickling W.G. 1993. The protective role of sparse vegetation in wind erosion. Progress on Physical Geography, 17, 50–68.

Woodruff N.P., Siddoway F.H. 1965. A wind erosion equation. Soil Science Society of America Proceedings, 29, 602–608.

Wu Zheng. 1987. Aaeolian Geomorphology. Beijing: Science Press. p. 153–166, 168–175. (In Chinese).

Wu Zheng. 2003. Sand landform and sand engineering. Beijing: Science Press, p. 61–65, 82–86, 93–97. (In Chinese).

Xiao Honglang, Li Xinrong, Duan Zhenghu, Li Tao, Li Shouzhong. 2003. Soil-vegetation system evolution in the process of sandy soil. Journal of Desert Research, 23(6), 605–611. (In Chinese).

Xu Shuying, Xu Defu. 1983. Aeolian sand accumulation in east bank of Qinghai Lake. Journal of Desert Research, 3(3), 11–17. (In Chinese).

Yang Hongxiao, Lu Qi, Wu Bo, Zhang Jintun, Sun Defu. 2006. Ecological restoration in Alpine sandy lands of Gonghe basin, Qinghai province. Science of Soil and Water Conservation, 4(2), 7–12. (In Chinese).

Yang Junping, Zhou Liye. 2000. China desertification situation and countermeasures research. Resources and Environment in Arid Areas, 14(3), 15–23. (In Chinese).

Yang Xiuchun, Yan Ping, Liu Lianyou. 2003. Advances and commentaries on wind erosion of soil. Agricultural Research in the Arid Areas, 21(4), 147–153. (In Chinese).

Yang Xueyang. 2010. We are repeating the same historical tragedy. GuangMing online, (2010-3-24)[2011-3-27]. http://www.360doc.com/content/10/0327/22/108917220528694. shtml. (In Chinese)

Yao Honglin, Yan Deren, Hu Xiaolong, Liu Yongjun, Zhang Huazhen. 2001. Research on the law of wind erosion and deposition of sediment in moving sand dunes in Maowusu sandy land. Mongolia Forestry Science and Technology, (1), 3–9. (In Chinese).

Ye Qingchao. 1992. Law of Surface Material Migration and Geomorphology in Yellow River Watershed. Beijing: Geology Publishing House, p. 31–40. (In Chinese)

Zhang Caiqin, Yang Chi. 2006. Simulation of plant growth and mathematical modeling study. Acta Scientiarum Naturalium Universities Neimongol, 37(4), 435–440.

Zhang Chunlai, Zhou Xueyong, Dong Guangrong, Liu Yuzhang. 2003. Wind tunnel studies on influences of vegetation on soil wind erosion. Journal of Soil and Water Conservation, 17(3), 31–33. (In Chinese).

Zhang Denshan, Shi Mengqi, Yang Henghua, Yang Hongwen. 2003. Rapid governance of moving sand dune in Qinghai-Tibet Plateau. Proceedings 2003 Academic Meeting of China Geographical Society. (In Chinese).

Zhang Dengshan. 2005. Conservation and reasonable utilization of desertification land in Sanjiangyuan. Proceedings, Qinghai-Tibet Plateau Environment and Change Seminar. Guilin. (In Chinese).

Zhang Hongjun, Liu Zhiguo, Gong Heping. 2009. Spatiotemporal change of soil water in fixed sand dune. Journal of Inner Mongolia Forestry Science and Technology, 35(1), 23–26. (In Chinese).

Zhang Hua, Li Fengrui, Fu Qianke, Lv Zijun. 2004. Field investigation on ecological effect of windbreak and soil erosion reduction from sandy grasslands. Environmental Science, 25(2), 119–124. (In Chinese).

Zhang Yinping, Zhang Jitao. 2000. The permanence of nonperiodic predator-prey system of three species Lotka-Volterra. Applied Mathematics and Mechanics, 21(8), 792–797. (In Chinese).

Zhang Yongmin, Zhao Shidong. 2008. The present status, future scenarios and countermeasures of global desertification. Advance in Earth Sciences, 23(3), 306–311. (In Chinese).

Zhao Chunhun. 2005. Experts say there is a large freshwater lake in Lop Nor thousands of years ago. Xinhua Network, (2005-03-27)[2011-03-27]. http://news.tom.com/1002/3291/200532 7-1988910.html. (In Chinese).

Zhao Cunyu, Wang Tao. 1999. The research status and prospect on vegetation succession in the process of sandy desertification grassland. Forest Sciences, 35(3), 103–108. (In Chinese).

Zhao Xiufeng. 1991. The cause and its environmental significance of the wind sand accumulation along Qinghai-Tibet highway. Resources and Environment in Arid Areas, 5(4), 61–69.

Zhou Huarong, Wang Lianshe, Li Xinhua. 2002. Preliminary experiment results of soil wind erosion in Yancao basin. Journal of Soil and Water Conservation, 16(6), 26–27.

Chapter 6

Erosion and vegetation

6.1 EROSION IN THE YARLUNG TSANGPO BASIN

6.1.1 Various types of erosion

According to the Columbia Encyclopedia (Columbia University, 2000), erosion is generally defined as the processes by which the surface of the earth is constantly being worn away. In other words erosion means the detachment and removal of solid particles from their original location on the landscape. Weathering is defined as the process of chemical or physical breakdown of the minerals in rocks (Halsey et al., 1998). Erosion and weathering may occur concurrently and weathering can be regarded as a part of erosion. Erosion can be classified according to the principal agents into gravity erosion, glacier erosion, water erosion, and wind erosion. Gravity erosion mainly refers to avalanches and landslides. Water erosion can be further classified according to its forms into splash erosion, sheet erosion, rill erosion, gully erosion, and channel erosion. In the Yarlung Tsangpo basin numerous debris flow events have occurred, which have moved a huge amount of moraine deposits a distance of several kilometers. Thus, debris flow erosion is also regarded as one of the main types of erosion in the Yarlung Tsangpo basin.

Various types of erosion occur in the Yarlung Tsangpo basin. In general, wind erosion and water erosion occur in the upper reaches. Gravity erosion occurs in the lower reaches (e.g. the Yarlung Tsangpo Grand Canyon). Glacier erosion occurs in the highlands of the eastern part of the basin and debris flow erosion occurs in the low lands of the eastern part of the basin.

The upper reaches of the Yarlung Tsangpo River from its origin to the confluence with the Lhasa River is a dry valley (Fig. 6.1), which has an average elevation of 4,500 m, an annual average temperature of $-0.3\sim1.2°C$, and an annual precipitation of 136–290 mm (Sun et al., 2010). Wind erosion and water erosion are the main types of erosion in the upper reaches (Zhong et al., 2005). Rainstorms occur in the wet season and cause splash erosion, rill erosion, and gully erosion. The river flow sorts the sediment in the valley and very strong wind in spring initiates movement of the fine sand from the floodplain and develops aeolian sand dunes on the bank slopes (Miehe et al., 2006). In the dry valley, a study of the erosion rate via a cesium 137 (^{137}Cs) inventory indicated that the vegetation coverage is the most important control factor for water erosion and wind erosion (Wen et al., 2000).

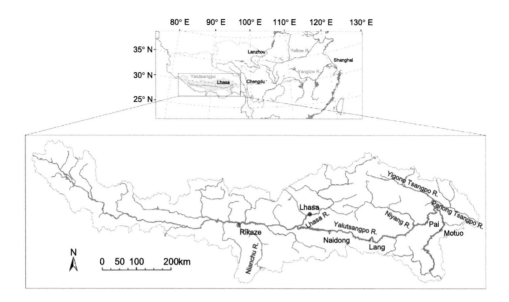

Figure 6.1   The Yarlung Tsangpo River and its tributaries.

The lower reaches are downstream from the Pai Town and mainly comprise the Yarlung Tsangpo Grand Canyon. The elevation of the river sharply decreases from 3000 to 200 m. The temperature and precipitation also increase sharply. For instance, the average temperature is 16°C and the precipitation is 2300 mm at Motuo (Hou and Wang, 2009). The great difference in elevation, temperature, and precipitation; complex terrains; and moisture inflow from the Indian Ocean along the canyon have resulted in various types of vegetation and cause different types of erosion.

The middle reaches, where the elevation reduces from 3600 to 3000 m, are between the upper and lower reaches. The average temperature and precipitation vary slightly in this region. At Naidong the annual average temperature is 7.5°C and the average precipitation is 443.6 mm (Lu, 2008). Vegetation develops quite well in the middle reaches region. Water erosion is the main type of erosion, although sand dunes occur in the valley around the confluence with the Niyang River.

Erosion and vegetation were measured and sampled almost every year by field investigations in the Yarlung Tsangpo River and the tributary rivers of Nianchu, Lhasa, Niyang, Parlong Tsangpo, and Yigong Tsangpo from 2009 to 2014. Table 6.1 lists the environmental factors of the sample plots. The sample plots were positioned by a Magellan eXplorist 210 GPS (Global Positioning System). Several methods were used to calculate or estimate the erosion rate at the sampling sites. A Trupulse 360 laser ranger was used for distance measurement, the maximum range of which was up to 2000 m with an accuracy of 1 m for distance, 0.25 degree for inclination, and 1 degree for azimuth. Shuttle Radar Topography Mission (SRTM3) data, with a resolution of 90 m, from the U.S. National Aeronautics and Space Administration (NASA) provided the basic digital elevation model (DEM) of terrain data. The slope angle and aspect

Table 6.1 Environmental factors of sediment and vegetation sample plots.

Plots	Location	Slope (°)	Aspect	Elevation (m)	Annual precipitation (mm)	Lithology and soil
S1	N29°49′33.4″, E92°21′19.7″, Milha Moutain	22	SE162	4969	250	Moderate acid tuff and volcanic breccias. Freeze-thaw weathering soil.
S2	N29°42′32.3″, E92°17′55.5″, Lhasa River valley	25	SE112	4515	545	Granite. Slope deposit.
S3	N29°45′38.1″, E91°55′39.7″, Lhasa River valley	23	NE150	3965	474	Slate, fine sandstone, limestone, quartz vein. Slope deposit.
S4	N29°35′44″, E90°59′32″, Lhasa River valley	28	SE125	3650	363	Slate, phylite. Slope deposit.
S5	N29°28′25.7″, E90°55′58.9″, Lhasa River valley	26	NW330	3618	363	Granite. Slope deposit.
S6	N29°14′41.5″, E91°40′48.3″, Yarlung Tsangpo valley	15	NW337	3588	423	Granite, granodiorite, and sandstone. Slope deposit and eolian sediment.
S7a	N29°34′7.6″, E91°0′21.7″,	0	–	3609	363	River sediment: gravel and sand
S7b	Sand bar in the Lhasa River					
S8	N29°34′19.3″, E90°59′19.1″, Sand dunes, Yarlung Tsangpo valley	0–30	–	3720	363	Eolian deposit: fine sand
S9	N30°1′31.9″, E95°0′150.1″, Parlong Tsangpo River	18	NW356	1971	1100	Quartz schist, biotite schist, biotite-plagioclase gneiss, and quartzite. Slope deposit
S10	N29°6′29.6″, E93°27′4.7″, Yarlung Tsangpo River at Milin.	24	NW310	2993	705	Granite. Eolian deposit and slope deposit

direction were measured with the laser range meters. The rocks in the sample plots were sampled for lithological analysis. Soil and sediment deposits resulting from different types of erosion were sampled for size distribution analysis. Cobbles and boulders were measured with scales and photographic analyses.

Wind erosion is a main type of erosion in the dry valley region in the upper reaches. The Parlung Tsangpo River transports a lot of sediment and sediment deposition in the wide section of the Yarlung Tsangpo River forms many sand bars (Fig. 6.2(a)). Sediment in the sand bars consists mainly of fine sand that is liable to be blown away in the dry season from January to May. Fine sand was lifted by strong winds and

Figure 6.2 (a) Sand bars consisting mainly of fine sand in the Yarlung Tsangpo River; (b) Eolian sand dunes on the bank slope of Yarlung Tsangpo valley near the Lhasa Airport.

then deposited on hill bank slopes. The resulting eolian sand dunes occur in the wide valleys of the Yarlung Tsangpo, Lhasa, and Nianchu rivers. Barchan dunes, crescent dunes and dune chains were observed in these dry valleys. Figure 2(b) shows sand dunes on the hill slope of the Yarlung Tsangpo valley near the Lhasa Airport. Sand dunes buried the original dry valley vegetation and the movement of sand dunes killed plants. Afterwards, very poor sand dune vegetation slowly developed if the dunes were not so actively movable. Only a few pioneer species could colonize the sand dunes.

Vegetation, although poor, develops on some of the sand dunes. A few tree and shrub species grow on the sand dunes and they suffer from wind erosion. At a few places the roots of shrubs and trees were exposed due to the wind erosion. The erosion time, T, is the age of the exposed roots. The thickness of the eroded sand layer, H, was directly measured from the collar of the shrub or tree to the present ground surface. Then the rate of Aeolian erosion, E_r, is given by:

$$E_r = 1000000 \frac{m^2}{km^2} \frac{\gamma_s H}{T} \tag{6.1}$$

where γ_s is the specific weight of sand. This method was used to get the erosion rate for the area to which S8 and S10 (Table 6.1) belong. The measured rate of erosion is in the range of 50,000–200,000 t/km²/yr, and the average value is 100,000 t/km²/yr.

In should be pointed out that: 1) different from other types of erosion wind erosion has no fixed transportation direction or destination. For a given location where wind erosion may have occurred in one period and eolian sand deposition and accumulation may occur in another period. Therefore, the rate of erosion does not reflect a long term continuous loss of sediment from the area but just the average intensity of erosion over time; 2) every year a small portion of fine sand was transported by wind over the hill slope and deposited on the Qinghai-Tibet Plateau. Overtime a huge amount of eolian sand accumulated at an area on the plateau causing desertification. Fortunately, most

Figure 6.3 Rill erosion on the dry valley banks (left) and gully erosion in the middle reaches of the Yarlung Tsangpo basin (right).

of the eolian sand accumulated over the past million years on the plateau has been stabilized by alpine steppe vegetaion, alpine meadow, or alpine shrubs, which have developed in wet years. Below the top vegetation layer there is a huge amount of eolian sand. Any demage to the vegetation may cause liberation of a great amount of eolian sand and desertification of the plateau, which is discussed in Chapter 5.

Water erosion occurs in the dry valleys of the upper Yarlung Tsangpo basin having main forms of rill erosion and gully erosion. As shown in Fig. 6.3, rills cut the dry valley banks. The number, length, width and depth of rills were measured and the time of erosion was estimated from the growth rings of the shrub stems on the gully bank and the exposed roots of the shrubs. In general the rills and gullies are 100–1000 m long, 0.2–2 m deep, and 0.5–2 m wide. By counting the number of rills and measuring the length, width, and depth with the laser range meter and scales the volume of erosion was estimated. The time of erosion was estimated by measuring the growth rings of the shrub stems on the rill and gully banks. The rate of erosion was obtained by dividing the erosion volume over the years of erosion, which was about 2000–5000 t/km^{2}yr.

Water erosion occurs in the middle reaches with the main form being gully erosion. The precipitation is moderate and the vegetation is much better than the dry valleys region. During rainstorms surface runoff erodes the banks and develops gullies, as shown in Fig. 6.3. The gullies are widened as a result of gully erosion. In general, the depth of the gullies does not change obviously. The rate of erosion is estimated using the following formula:

$$E_r = \frac{n\gamma_s Lh\Delta W}{AT} \tag{6.2}$$

where L and h are the average length and depth of the gullies, respectively; n is the number of gullies in the erosion area A; ΔW is the average value of the width change of the gullies due to gully erosion in the time period T.

In the Yarlung Tsangpo Grand Canyon and tributaries flowing into the canyon, avalanches occur very often because the riverbed has been experiencing continuous

Figure 6.4 A recent avalanche in the Yarlung Tsangpo Grand Canyon (left); two avalanche deposit fans on the Parlong Tsangpo River (right).

incision in the Pleistocene and Holocene (Finnegan et al., 2008; Zhao and Li, 2008). Figure 6.4 shows an avalanche at the Yarlung Tsangpo Grand Canyon turn (left) and two avalanches on the Parlong Tsangpo River, which is the largest tributary and flows into the Yarlung Tsangpo at the middle of the Grand Canyon (right). The essential cause of the avalanches was continuous river bed incision in the Grand Canyon and tributaries, although the direct cause of the events was rainstorms or earthquakes. The relative height from riverbed of the Yarlung Tsangpo or Parlong Tsangpo rivers below the mountains is 2,000–6,000 m. The river valley becomes super-V shaped, with an average bank slope higher than 30° with the lower part of the bank slope higher than 40°. The potential energy of the avalanches was very high.

The erosion rate was estimated by measuring the volume of avalanches along the river and by estimating the time of the events. Twenty eight avalanches were measured along a 23 km long reach of the Parlong Tsangpo River in the past 10 years. There are almost no trees or shrubs but only very poor herbaceous vegetation on the new avalanche deposits. The area is wet and warm and vegetation develops quickly. The time of the avalanche events was estimated as no longer than 10 years. Thus, the erosion rate was estimated at 90,000 t/km²yr. This method was used to get the erosion rate for the area to which S9 (Table 6.1) belongs.

River bed incision also caused many landslides. A huge landslide with a volume of 500 million m³ occurred on Zhamunon Creek in 1900. The landslide dammed the Yigong Tsangpo River and created Yigong Lake and a 100 m high knickpoint on the river. A century later another huge landslide occurred on Zhamunon Creek in 2000. A landslide with a volume of 300 million m³ slid 8.5 km and dammed the Yigong Tsangpo River again. The landslide dam enlarged Yigong Lake after a partial failure. Figure 6.5 shows the landslide from Zhamunon Creek and the residual of the failed landslide dam. The catchment area of the Zhamunon Creek is 22.5 km², which is calculated with a 30 m accuracy digital elevation map. Assuming the reoccurrence period of landslides is 100 years, the average rate of landslide erosion is 200,000 t/km²/yr, in which the density of the landslide body is estimated as 1.5 t/m³.

Figure 6.5 A landslide from Zhamunon Creek with a volume of 300 million m³ occurred on April 09, 2000 (left) and the residual of the failed landslide dam on the Yigong Tsangpo River (right).

Landslides, avalanches, and glacial moraines provide plenty of loose solid materials for debris flows (Burbank et al., 1996). Debris flow erosion is a special type of erosion, which cannot be included into gravity erosion or water erosion. Moreover, debris flow erosion is also a main type of erosion in the area (Garzanti et al., 2004; Shang et al., 2003). Debris flows occur very often in the basin of the Yarlung Tsangpo Grand Canyon and Parlong Tsangpo River. Large debris flow events were investigated and the volume of solid materials transported by debris flows have been estimated and reported since 1950. The debris flow gully basins were investigated and the sediment from these basins was sampled. The rate of debris flow erosion is estimated from the total volume of debris flow and basin area.

The Guxiang Gully is a tributary of the Parlong Tsangpo River, where debris flows occur every year. Debris flow scours avalanche and moraine deposits in the gully and transports the sediment into the Parlong Tsangpo River. A huge debris flow occurred in the gully on September 29, 1953, which transported 9 million m³ of sediment into the Parlong Tsangpo River. A huge stone weighing 4000 t was transported to the fan by the debris flow. From then on almost every year several debris flows occurred and the annual sediment volume transported by the debris flows was more than 1 million m³. Figure 6.6 shows the Guxiang Gully and many avalanche deposit fans (left). It is estimated that there are about 400 million m³ of loose solid materials in the gully, which are mainly from glacial moraine, avalanches, and landslides deposits. The outlet of the gully is a gorge section with a rock bed. Debris flow carried cobbles and gravel and abraded the bedrock forming a 12 m deep, 6 m wide bedrock channel (right).

In the past 60 years more than 100 million m³ of sediment had been eroded from the Guxiang Gully. Figure 6.7 shows the variation of a cross section of the Guxiang Gully from 1954 to 1994 due to debris flow erosion, in which *H* is relative height. From 1954 to 1965 several debris flow events occurred every year, which scoured the gully at an average rate of 3.3 million m³ per year. From 1965 to 1973 the average rate of debris flow erosion was 0.9 million m³ per year. From the 1990 to 2005 the

Figure 6.6   The Guxiang Gully and many avalanche deposit fans (left); and a 12 m deep, 6 m wide bedrock channel at the outlet of the Guxiang Gully (right).

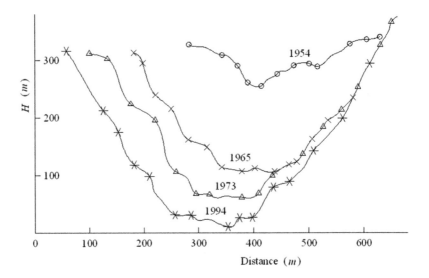

Figure 6.7   Variation of a cross section of the Guxiang Gully from 1954 to 1994 resulting from debris flow erosion.

debris flow gully was relatively quiet and the rate of debris flow erosion reduced to 0.1 million m³/yr. In the recent decade debris flows have not occurred every year and the average rate of erosion was less than 0.05 million m³.

Debris flows transported sediment into the Parlong Tsangpo River and dammed the river several times. The debris flow dams partially failed leaving residual boulders, which composed energy dissipation structures and formed a 200 m high knickpoint on the river, as shown in Fig. 6.8. Guxiang Lake was formed by a debris flow dam, which

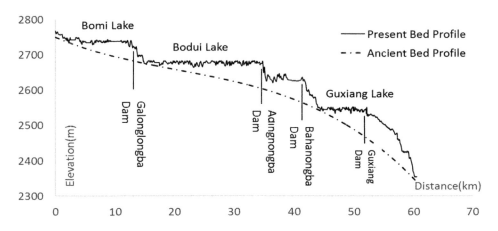

Figure 6.8 Longitudinal bed profile of the Parlong Tsangpo River showing the knickpoint formed by a debris flow dam at the confluence of the Guxiang Gully and Guxiang Lake.

has been filled with sediment carried by the river flow from upstream. Suspended load and bed load deposited upstream of the dam and formed gravel and sand bars, which had a rather uniform size distribution. Galonglongba Dam shown in Fig. 6.8 also is a debris flow dam, which partially failed and the residual material forms another knickpoint on the Parlong Tsangpo River.

Figure 6.9 shows the particle size distribution of sediment of different erosion types, in which a debris flow deposit was sampled at the mouth of the Guxiang debris flow gully; a landslide deposit was sampled at the remaining part of the Yigong Landslide Dam on the Yigong Tsangpo River; a glacial deposit was sampled at the Midui Glacier Moraine Fan; a dry valley deposit was sampled on the slopes of the Lhasa River valley; a sediment bar deposit was sampled at an upstream sediment bar of the Guxiang debris flow dam; the soil of Milha Mountain was sampled at the slope of Milha Mountain; eolian sand was sampled at the eolian deposit on the slope of the Lhasa Valley; and the sediment of a sand bar was sampled at a sediment bar in the Lhasa River.

Debris flows, landslide, and glacial deposits, and the slope deposit on the bank slope of the dry valleys region had the widest size distribution. The coarse part of the sediment could be transported for a short distance by the river flow as bed load. Only less than 10% of the fine sediment can be transported for a long distance to India as suspended load. The eolian sand dunes and the surface sediment of bars in the Lhasa River had similar size distributions because the eolian sand dune was composed of the fine sand from the river sediment.

6.1.2 Glacier erosion

There are more than 4000 glaciers in the Yarlung Tsangpo Basin, among them 1,800 glaciers are in the tributary Parlong Tsangpo basin. Glaciers are very active and glacier erosion is the most important type of the erosion in the high mountains around the Yarlung Tsangpo Grand Canyon. There are two types of glaciers in the area: hanging

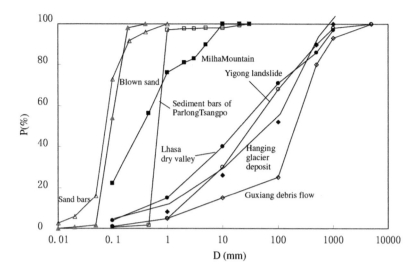

Figure 6.9 Grain size distributions of various types of erosion (*D* is the diameter of sediment; *P* is the percentage of particles finer than the corresponding *D*).

glaciers and valley glaciers. Hanging glaciers transport sediment directly into rivers and forms moraine fans on the river, as shown in Fig. 6.10. In general hanging glaciers cut the gully bed deep and narrow. The bed gradient is high and uniform, generally within the range of 30°–35°. Hanging glaciers are generally several hundreds of meters long and have ice depth of only a few meters. Therefore, such glaciers are not able to cut huge rocks. The sediment carried by hanging glaciers is relatively fine, with a median diameter around 0.1 m (Fig. 6.9). Sediment transported into the river by hanging glaciers can be carried downstream by flood water. Thus, hanging glaciers cannot dam the river and do not substantially affect the fluvial form of the river.

The rate of hanging glacier erosion was estimated by counting the number of glaciers, measuring the annual movement distance of the glaciers, and sampling the sediment in the ice. Thus, the rate of glacier erosion is given by:

$$E_r = \frac{SLBH}{A} \tag{6.3}$$

in which S is the sediment concentration in the ice by weight; L is the average annual movement distance of the glacier, B is the average width of the glacier, H is the average thickness of moving ice, and A is the drainage area of the glacier. The movement distance of the glacier was calculated by locations of the downstream end of the glacier in summer and winter or spring. In general the downstream end of the glacier in winter is located at the river. In some cases, the glaciers may move across the frozen river and arrive at the opposite bank. Therefore, the annual movement distance of hanging glaciers is the distance from the tongue of the glacier in summer to the river. In the upper reaches of the Parlong Tsangpo River there are several tens of hanging glaciers. The movement distance of the hanging glaciers is 100 to 300 m. The concentration of

Figure 6.10 A hanging glacier (left) transports sediment into the Parlong Tsangpo River and forms a moraine fan on the river (right).

Figure 6.11 (a) Tianmo Gully Glacier cut the valley wall and caused a rock avalanche; (b) A cutting gap near the glacier wall and striations formed by large shear force of the glacier.

solid materials in the hanging glaciers was measured as 5–28 kg/m^3. The rate of glacier erosion calculated with Eq. (6.3) was 2,000–50,000 t/km^2/yr for the hanging glaciers.

Valley glaciers are much larger in size than hanging glaciers. Valley glaciers transport a huge amount of moraine sediment and deposit it in the valley, and the sediment can be transported for a distance into rivers by debris flows. The width of valley glaciers is 100–500 m and the thickness of the ice is 3–20 m. Huge valley glaciers have great power to carry sediment and cut the rock valley bed and walls. Figure 6.11 shows the Tianmo Gully Glacier, which cut the valley wall and caused two rock avalanches. There is a curved cutting gap near the glacier wall, which is formed as the glacier moved down and cut the valley wall. The sediment eroded by valley glaciers is coarser than that of hanging glaciers. The sediment is produced in two processes: 1) moving ice cuts the bed and valley walls yielding boulders; 2) snow and ice carry relatively fine solid particles resulting from freeze–thaw erosion down the slope on to the glacier surface. Therefore, the solid materials from glacier erosion have two dominant groups: boulders of diameter around 100–1000 mm and gravel of diameter around 1–10 mm.

The rate of glacier erosion is very different for different valley glaciers. Galongla No. 3 Glacier is an active valley glacier and the erosion rate is relatively high. The volume variation of the moraine channel section was measured at 14 cross sections from 2012 to 2014. The erosion rate of the valley glacier can be calculated using the volume increase of the moraine deposit and is given by

$$E_r = \frac{V_{t2} - V_{t1}}{TA} \qquad (6.4)$$

in which V_{t2} and V_{t1} are the volumes of moraine deposits of the glacier at time t_2 and t_1, respectively; A is the drainage area of the glacier, and T is the time interval between the two measurements. For Galongla No. 3 Glacier, $A = 2.8\,km^2$ and the average erosion rate of the glacier is 200,000 t/km^{2}yr.

6.2 VEGETATION IN THE YARLUNG TSANGPO BASIN

6.2.1 Various types of vegetation

Various types of vegetation develop in the Yarlung Tsangpo Basin. Alpine steppe, alpine meadow, alpine shrub, and scree vegetation were reported in the upper reaches (He et al., 2005). The middle reaches of the river is in the dry valleys region. Dry valleys have two unique features, which may be used as diagnostic characteristics: 1) deeply incised valleys on a plateau; and 2) significantly higher temperatures and evaporation rates and lower precipitation than the surrounding area on the plateau. Poor vegetation develops in the dry valleys (Zhao et al., 2007; Zhao et al., 2006). From the middle reaches to the lower reaches in the east of the plateau transitional mixed coniferous or evergreen broad-leaved forest occurs (Zhang et al., 2008). Nevertheless, in the Yarlung Tsangpo Grand Canyon subtropical vegetation occurs consisting of evergreen broad-leaved forest, mountain rainforest, and monsoon forest (Sun et al., 1997; Zhen, 1999).

Vegetation can effectively control water erosion and wind erosion. Vegetation reduces erosion by 1) adsorbing the impact of raindrops, 2) reducing the velocity and scouring power of runoff, 3) reducing the runoff volume by increasing the percolation into the soil, 4) binding soil with roots, and 5) protecting the soil from wind erosion (Goldman et al., 1986). A study indicates that the soil erosion on the Loess Plateau is inversely proportional to the density of the vegetation cover. The sediment yield reduces to nearly zero if the forest cover is higher than 60% (Wang and Wang, 1999). For a watershed, vegetation and erosion may reach an equilibrium state if the circumstances remain unchanged for a long period of time. However, the equilibrium is not stable. Ecological stresses, especially human activities, may disturb the balance and initiate a new cycle of dynamic processes. The first author developed the so-called vegetation-erosion dynamics, which can be applied to evaluate the evolution of the vegetation under the action of erosion and various ecological stresses (Wang et al., 2004; Wang et al., 2008).

Vegetation and sediment sampling and erosion measurement were conducted at 10 sampling plots (Table 6.2). The ten vegetation plots, $10 \times 10\,m^2$ each, were selected with typical vegetation types at different elevations and for different erosion types. Specimens of all plant species were collected and photographed for identification. Numbers of species were counted. Average height and canopy width in two mutually perpendicular directions (for coverage calculation) were measured.

There is no vegetation in the mountains higher than 5200 m in the Himalaya and Transhimalaya mountains because of perennial ice cover. An alpine mat vegetation develops on the mountains at elevations between 4000 and 5000 m, with a thickness of only 1–5 cm. The vegetation consists of scrubby herb and bryophytes species. The coverage of the vegetation is more than 90%. Although the ground height of the vegetation is less than 5 cm the underground root network is dense and extends to a depth of 60 cm.

In the middle reaches of Yarlung Tsangpo River, drought tolerant species dominate the poor dry valley vegetation on the bank slopes. Numerous sandbars are present in the wide valleys. Sandbar vegetation colonizes on the unstable bars and islands. In the Yarlung Tsangpo Grand Canyon, however, subtropical vegetation and coniferous vegetation develop. Between the middle reaches and the Grand Canyon transitional vegetation types develop, which is mainly composed of temperate shrubbery. Table 6.2 lists the type of vegetation and species composition.

Figure 6.12 shows the alpine mat vegetation at Milha Mountain (4960 m). The thickness of the vegetation was only about 2 cm. The vegetation had a dense and deep root system. There were more than 10,000 roots per m^2 and the dense roots extended to 60 cm below the ground. Hard low herbaceous plants, mosses, and lichens dominated the vegetation. *Soeoserisgillii, Gentianana sp., Androsacebrachystegia, Aphragmusoxycarpus, Sedum sp.* and *Draba sp.* composed the plant community. The vegetation had relatively high species richness. The dense root system effectively protects the soil against erosion.

The dry valley vegetation occurs mainly on the hill slopes of the YarlungTsangpo valley and Lhasa and Nianchu river valleys at elevations of 3500–4000 m. The annual precipitation is 300–500 mm and the annual evaporation is as high as 2688 mm in the valleys. Only sun plants and mosses grow, which are accustomed to the low precipitation and high evaporation, high sun light, and thin soil layer. The species diversity is low and vegetation coverage barely reaches 10%.

Figure 6.13 shows typical dry valley vegetation on the bank slope of the Lhasa River and moss and herbaceous species on a granite rock surface. The lithology has influenced the species composition. Only mosses and lichens grow on the surface of granite rocks. Drought tolerant herbaceous species, like *Artemisia vestita, Ceratostigmag riffithii,* and *Pteris tibetica,* grew in weathering crevices and soil. More species have colonizedin metamorphic bedrock areas. *Sageretia gracilis, Rhododendron nivale, Thalictrum diffusiflorum, Meconopsis simplicifolia, Clematis tenuifolia, Aconitum richardsonianum, Artemisia viscidissima,* and *Potentilla fruticosa* have been found on the metamorphic rock slope of the Lhasa River. Mosses and lichens protect the rock from weathering, especially for granite which has coarse mineral crystals. In fact granite with mosses and lichens covering its surface had much less weathering than that without mosses and lichens.

Table 6.2 Types of vegetation and species composition at 10 sampling plots in the Yarlung Tsangpo basin.

Plots	Vegetation	Species
S1	Alpine mat vegetation	Soeoseris gillii, Gentianana sp., Cyperacea sp.(2), Androsace brachystegia, Delphinium sp., Saussurea sp., Aphragmus oxycarpus, Polygonum sp., Rhododendron nivale, Cyananthus sp., Sedum sp., Arenaria sp., Carex sp., Anemone imbricatum, Kobresia sp., Draba sp., Ranunculus sp., Saxifrag sp., Leontopodium sp., lichen and moss.
S2	Dry valley vegetation	Berberis hemsleyana, Myricaria prostrate, Potentilla parvifolia, Stipa purpurea Griseb., Pennisetum flaccidum, moss.
S3		Festuca ovina, Artemisia vestita, Berberis hemsleyana, Sageretia gracilis, Rhododendron nivale, Thalictrum diffusiflorum, Stellera chamaejasme, Cynodon dactylon, Meconopsis simplicifolia, Clematis tenuifolia, Aconitum richardsonianum, Artemisia viscidissima, Potentillafruticosa, Taraxacum sp., et al.
S4		Artemisia desertorum, Sophora moorcroftiana, Eragrostis pilosa, Setaria viridis, Festuca ovina, lichen and moss
S5		Setaria viridi, Artemisia vestita, Ceratostigmagriffithii, Stipabungeana, Myricariaelegans, Pteristibetica, Aleuritopterisargentea, lichen and moss
S6		Loniceraspinosa, Ceratostigma minus, Astragalusstrictus, Aristidateangpoensis, Ixerischinense ssp. Versicolor, Sophoramoorcroftiana
S7a	Sand bar vegetation	Pennisetum flaccidum, Hippophaerhamnoides, Myricariawardii, Salsolaruthenica, Salsolaruthenica, Artemisia younghusbandii, Artemisia carvifolia, Nitrariatangutorum
S7b		Salix variegate, Triticumaestivum, Hordeum vulgare var. coeleste, Brassica chinensis var. oleifera, Festucaovina, et al.
S8	Eolian sand dune vegetation	Artemisia desertorum, Populus simonii, Artemisia vestita, Artemisia xigazeensis, Asteraceae sp., Corispermum lepidocarpum, Oxytropis sericopetala, et al.
S9	Sub-tropical vegetation	Alnusnepalensis, Hydrangea sp., Maesacavinervis, Juglanssigillata, Debrgeasia edulis, Diandranthustibeeiticus, Boehmeriabicuspis, Impatiens arguta, Salix sp., Artemisia tangutiaca, Athyriaceae sp., Brassaiopsis sp., Aralia tibetica, Clematis buchananiana, Lepisorusscolopendrium, Rubiamanjith, Viola sp., Deutzia corymbosa, Gramineae sp.(2), Schizopeponbomiensis, Elatostemacuneiforme, Rubusmacilentus, Rubustreutleri, Tetrastigmaplanicaule, Rubusfohaceistipulatus, Rubusfohaceistipulatus, Neilliathyrsiflora, Clematis connate, Trichosanthesrubriflos, Hippochaeteramosissima, Cirsiuminterpositum, Dryopteris sp., Galiumasperuloides, Coniogramme intermedia Hieron. Var. glabra Ching, Hypericumuralum, Buddleja candida, Duchesneaindica, Periplocaforrestii, Stellaria sp.
S10	Transitional vegetation	Amygdalus mira, Asteraceae sp., Cotoneaster hebephyllus, Drynaria sinica, Rabdosia rugosa, Cotoneaster sherriffi, Rhamnus vigata, Rose sericea f. glandulosa, Thalictrum sp., Aristida triseta, Jasminum stephanense, Rabdosia parvifolia, Phlomis medicinalis, Pedicularis fletcherii, Cheilosoria insignis, Leguminosae sp., Coeallodiscus flabellatus var. leiocalyx, Clematis Montana, Phtheirospermum tenuisectum, Nothosmyrnium xizangense, Gymnopteris vestita, Artemisia vestita

Sandbar vegetation occurs on sand bars or islands in the Yarlung Tsangpo and Lhasa rivers, at elevations of 3000–4000 m. The coverage varies because of fluvial processes and human activities. Newly formed bars are not stable and have very low vegetation coverage. On bars with a relatively low frequency of flood inundation a simple plant community has developed consisting of pioneer species of *Pennisetum flaccidum, Hippophae rhamnoides, Salsola ruthenica, Artemisia younghusbandii,* and *Nitraria tangutorum.* The vegetation coverage on the bars is less than 20%. Humans have planted willows and poplars and stabilized some large bars and islands, which

Figure 6.12 Alpine mat vegetation at Milha Mountain and its dense root network in the soil.

Figure 6.13 Dry valley vegetation on the bank slope of the Lhasa River and moss and herbaceous species on a granite rock surface.

provide conditions for establishment of more riparian species. Sediment on bars is well sorted. Fine sand was lifted up, transported, and deposited by wind in depressions of the bank slopes. Eolian sand dunes of a height of several tens of meters have developed on the bank slopes. These eolian dunes were not suitable for plant growth, except of wind resistant pioneer species, such as *Sophora moorcroftiana, Artemisia vestita, Artemisia xigazeensis,* and *Oxytropis sericopetala.* Very poor eolian sand dune vegetation has developed on the sand dunes by the Yarlung Tsangpo and Lhasa rivers with elevations of 3000–4000 m. The vegetation coverage was only 1–5% in general. Nevertheless, on a few semi-fixed dunes the highest vegetation coverage might reach 30%.

Subtropical vegetation occurs on the steep bank slopes of the Yarlung Tsangpo River downstream from the Pai Town and the Parlung Tsangpo River with elevations of 700–3000 m. Warm moist inflows from the Indian Ocean travel northward

Figure 6.14 Subtropical vegetation in the Yarlung Tsangpo Grand Canyon (left); and a dominant species *Musa balbisiana* (right).

in the Yarlung Tsangpo Grand Canyon. As a result the Grand Canyon and its surrounding areas are quite warm and humid. Subtropical vegetation has developed in the canyon and surrounding areas. The species diversity is high. Herbaceous, shrub and wood species colonize the habitat. Some wood species have developed into forests with heights of 20 m. *Alnus nepalensis, Hydrangea sp., Maesa cavinervis, Diandranthus tibeeiticus, Boehmeria bicuspis, Artemisia tangutiaca, Athyriaceae sp., Lepisorus scolopendrium, Schizopepon bomiensis, Trichosanthes rubriflos, Coniogramme intermedia Hieron. Var. glabra Ching, Buddleja candida,* and *Duchesnea indica* were observed in the canyon. Figure 6.14 shows subtropical vegetation and subtropical species in the Yarlung Tsangpo Grand Canyon.

More than 4,000 hanging glaciers and valley glaciers occur on high mountains around the Yarlung Tsangpo Grand Canyon and Parlung Tsangpo River. There is no vegetation on the moving ice and active moraine channel surface. On the relatively stable fans of hanging glaciers and quite stable outside slopes of the moraine levees an arbor-shrub-grass compound vegetation grows. The age of trees is different at different relative heights on the outside slope of the moraine levees. Figure 6.15 shows the vertical distribution of vegetation on the outside slope of moraine deposits. Dense vegetation with high trees grow at the lower part of the moraine deposit. The trees are mostly fir. Cores were taken and the age of trees was measured at different relative heights. The age of trees reduces from the lower part to upper part of the moraine deposit. Thus the age of different layers of moraine deposits can be obtained. Using this method the rate of erosion and speed of moraine deposition can be roughly estimated.

6.2.2 Effects of elevation and erosion on vegetation

The vegetation in the Yarlung Tsangpo basin varies regularly with the elevation. Figure 6.16 shows the thickness of vegetation, or the average height of vegetation,

Figure 6.15 A bird's eye view of the vegetation on a moraine deposit of Galongla No. 3 Glacier.

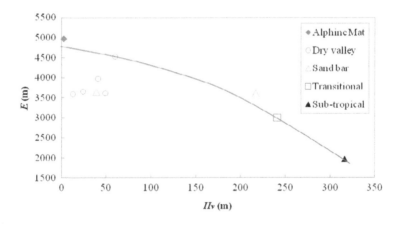

Figure 6.16 Average thickness of vegetation as a function of elevation in the Yarlung Tsangpo basin.

H_v, as a function of elevation above the sea level, E. The thickness was calculated as the sum of the height of plants weighted with its shadow area over the whole area of the sampling plot. The vegetation thickness was only 1–3 cm at an elevation of 5000 m, increased to 200 cm at about 3500 m and to more than 300 cm at 2000 min the Yarlung Tsangpo Grand Canyon.

Figure 6.17 shows the coverageor the area ratio of all plant cover over the whole area of the plot, V_{cover}, as a function of elevation. The coverage had the lowest value in the dry valleys at elevations of 3500–4000 m, which was generally less than 20%.

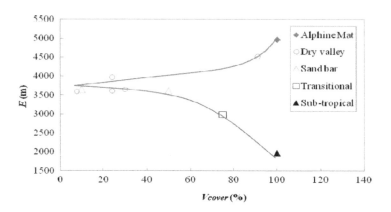

Figure 6.17 Vegetation coverage as a function of elevation in the Yarlung Tsangpo basin.

The precipitation increases from the dry valleys to the high mountains, with a gradient about 20 mm/100 m. The evaporation decreases with elevation because of temperature reduction. The coverage increased with elevation and reached nearly 100% at 5000 m. On the other hand the precipitation increases down the river from about 200 mm in the dry valleys to 3000 mm in the Yarlung Tsangpo Grand Canyon because of the moisture inflow from the Indian Ocean along the canyon. Therefore, the coverage increases as elevation decreased along the river valley from the dry valleys in the middle reaches to the Yarlung Tsangpo Grand Canyon. The coverage increased to almost 100% as the elevation decreased to about 2000 m in the canyon.

Figure 6.18 shows the species richness, or the number of species of plants, S, as a function of elevation. The species richness, S, was low, say less than 10 in general, in the dry valleys at elevations of 3500–4500 m. In the high mountains the species richness was not low because of the Alpine mat vegetation. If mosses and lichens were counted, the number of species would be more than 30. The highest species richness occurred in Yarlung Tsangpo Grand Canyon. More than 40 vascular species were found in a $10 \times 10\,m^2$ sampling plot.

The vegetation and erosion were mutually controlled while they adjusted to each other. Table 6.3 lists typical vegetation and erosion types and estimated annual rates of erosion. The rate of erosion was very different in the areas with different vegetation types. On Milha Mountain dense roots of alpine mat vegetation greatly reduced the erodibility of the soil (Zhang et al., 2007). The top soil was quite deep, with a depth larger than 1 m. The rate of erosion was nearly zero.

The middle reaches of Yarlung Tsangpo River have a lotus-root shape with wide valley sections of a width of 3–10 km between narrow gorge sections with widths of only a hundred meters. The lithology of bed rock by the river is mainly granite, granodiorite, and metamorphic rocks. Weathering and rock erosion occurred as a result of expansion and contraction due to temperature changes and breakdown of rocks under the action of sunshine. Without the protection of vegetation cover, bare rocks were acted on by the radiation from the sun and temperature changes. The exposure to weathering and the cycle of expansion during the day and contraction

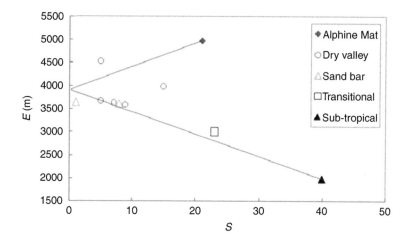

Figure 6.18 Species richness in the sample plots as a function of elevation.

Table 6.3 Erosion types and annual rate of erosion related to vegetation types.

Plots	Vegetation type	Coverage (%)	Elevation (m)	Soil composition	Erosion type	Estimated erosion rate (t/km²yr)
S2,S3, S4,S5, S6	Dry valley vegetation	5–15	3000–4000	Cobbles, gravel, sand, and a small amount of clay.	Rill erosion and gully erosion.	2000– 5000
S1	Alpine mat vegetation	80–95	4000–5200	Sand, clay, and gravel	Freeze-thaw erosion	Almost 0
S7	Sand bar vegetation	0–20 (may reach 60 for planted vegetation)	3000–4000	Sand	Channel erosion	0–10,000 depending on the vegetation
S8	Eolian sand dune vegetation	0–30	3000–4500	Fine sand	Eolian erosion	90,000
S9	Subtropical vegetation	80–99	1000–3000	Boulders, gravel, sand, and clay	Gravity erosion and debris flow erosion	50,000– 140,000
S10	Moraine vegetation	20–70	3000–4500	Gravel and sand	Glacier erosion	2000– 130,000

in the night caused fissures and produced gravel and sand (Wang et al., 2011). The loose solid materials were carried into the river by torrential floods. The sediment deposited in the wide valley sections and formed sand bars. Most of the sand bars were not stable and very poor vegetation developed. In the past half century humans have planted willow and poplar trees on some large bars and stabilized these bars. As a result, vegetation developed and the bars became stable islands with almost no erosion for these islands. On the other hand, some bars with no or poor vegetation were scoured and removed very quickly. Comparison of the locations of the bars in the period from 2000–2010 using satellite images finds that these bars were scoured away by the flood flow just in one year. The erosion rate of sand bars differs greatly depending on the human planted vegetation. The highest rate of erosion was larger than 10,000 t/km²/yr.

In the Yarlung Tsangpo Grand Canyon and lower Parlong Tsangpo River subtropical vegetation develops, which has high coverage and average height. There was no evidence of water erosion or wind erosion. Gravity erosion and glacier erosion occur in the area. These types of erosion mainly occurred in the lower Yarlung Tsangpo and Parlong Tsangpo basins and cannot be controlled by vegetation.

In summary, water erosion and wind erosion have occurred in the upper and middle reaches of the Yarlung Tsangpo basin. They have been highly affected by vegetation and also greatly affected the development of vegetation. Dry valley vegetation was poor, which consisted of sun species, mosses and lichens. Water erosion and wind erosion occurred in the dry valley region and the erosion rate was not high. Wind erosion was high, but only occurred in the area with sand dunes. Landslides and avalanches occurred in the lower Yarlung Tsangpo and Parlong Tsangpo rivers and the erosion rate was extremely high. Because the eroded materials were rather coarse, only a small portion of the sediment could be transported by river flow. Glacier erosion varied over a wide range from glacier to glacier. Gravity erosion, glacier erosion, and debris flow erosion cannot be controlled by vegetation.

6.3 GRAIN EROSION AND CONTROL STRATEGIES

6.3.1 Grain erosion

Grain erosion is a unique type of erosion that occurs intensively on the eastern margin of the Qinghai-Tibet Plateau, especially in the dry valleys in Yunnan and Sichuan provinces. Grain erosion is defined as the phenomenon of breaking down of bare rocks under the action of sun exposure, temperature changes, freeze-thaw cycle, detachment of grains by wind, flow of grains down the slope under the action of gravity, and accumulation of eroded materials at the toe of the mountain forming a deposit fan (Wang et al., 2010). The grains, 0.1–200 mm in diameter, jump and hit against the mountain slope surface and cause detachment of slope debris or rock surface, which results in further erosion. Grain erosion is an integration of weathering, shattering erosion, wind erosion, and gravity erosion, and thus cannot be classified into any erosion type related to a single agent. Grain erosion deposit fans look similar to shattering erosion fans but the intensity of grain erosion is 1000 times higher than shattering erosion. In the Jinsha River valley on the eastern margin of the Qinghai-Tibet Plateau it was observed that about 95,000 grains with an average diameter of 3 cm flowed down the slope per minute due to grain erosion. The largest diameter of the grains was about 20 cm. In general, grain flow occurs in the afternoon when it becomes windy. It was too dangerous to measure directly the erosion rate. Three video cameras were used to estimate the rate of erosion with video data. The total erosion volume from one grain erosion site was 1,800 m³ in one month (May 2011). The 'flood season' of grain erosion is from March to June with little grain erosion in other seasons.

(a)   (b)

Figure 6.19 (a) Freezing and thawing of rocks has resulted in shattering erosion on a 4700 m high mountain at Luhuo on the Qinghai-Tibet Plateau; (b) Eroded sediment resulting from shattering erosion deposits at the toe of the Rocky Mountains in Canada.

Shattering erosion can be regarded as a part of grain erosion. Shattering erosion of rocks and below stratified talus deposits has been reported from temperate upland environments as well as cold environments worldwide (Saas and Krautblatter, 2007). Stratified scree deposits with rich fine sediment are not only confined in the areas where a periglacial climate and prior glaciation existed or had once existed, but also occur in vegetated upland environments (Garcia-Ruiz et al., 2001). Usually shattering erosion is found in high mountains with elevations of 1200–4000 m (Matsuoka, 2008). But it is also found in south Walse at 650–770 m (Harris and Prick, 2000). The lithology of the weathering rocks can be limestone, dolostone, marls, or sandstone (Curry and Black, 2002).

Little data have been collected concerning the rate of shattering erosion because the rate is very low and difficult to measure. Thirteen years of observations in the southeastern Swiss Alps found that shattering erosion occurred at an average rate of about 0.1 mm/yr with significant spatial and inner-annual variations (Matsuoka, 2008). Figure 6.19 shows shattering erosion in the eastern margin of the Qinghai-Tibet Plateau and Rocky Mountains in Canada. The deposits are called stratified scree deposits or alpine scree slopes in some literature (Hetu et al., 1995). Chinese researchers have studied the structure of grain deposit fans, repose angle, and stability of the fans (Wang et al., 2007). These researchers focused on the velocity distribution of grain flow on the slope and the pressure exerted on the protection walls of highways by deposit fans. Other researchers have studied the critical initiation of slump of grain deposits on slopes (Chang et al., 2006).

Different from shattering erosion, grain erosion is a short term and very intensive erosion. Grain erosion occurs on fresh bare rocks resulting from rock falls, bank failures, avalanches, and landslides in the eastern margin of the Qinghai-Tibet Plateau.

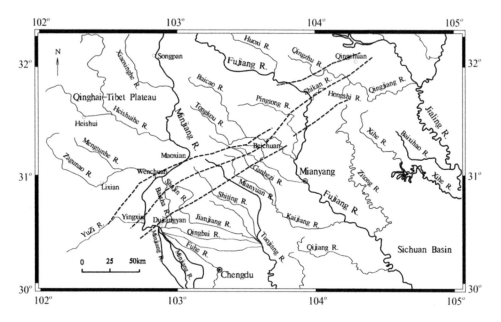

Figure 6.20 Longmenshan fracture belt consisting of the Yinxiu-Beichuan-Qingchuan fault, back fault and front fault, and streams from the Qinghai-Tibet Plateau to the Sichuan Basin.

Mass movements triggered by the Wenchuan Earthquake in 2008 have left a huge area of bare rocks (Wang et al., 2009). Erosion of rocks has been occurring in the form of grain detachment and movement. The erosion is extremely intense. The maximum rate of grain erosion after the Wenchuan Earthquake was measured as 50 cm/yr, which is equal to 1.35 million t/km^{2}yr.

The Wenchuan Earthquake, which occurred on May 12, 2008, caused many huge landslides and avalanches in many river valleys on the eastern margin of the Qinghai-Tibet Plateau. In fact, the earthquake occurred as a result of tectonic motion of the Qinghai-Tibet Plateau along the Yingxiu-Beichuan fault, which is the central fault of the Longmenshan fractural belt (Fig. 6.20). There are many rivers, which deeply cut the mountains, in the quake-hit area.

Several thousands of avalanches and landslides were triggered by the earthquake, which have left a huge area of bare rocks. Grain erosion has been occurring on the fresh bare rock surface. The bare rocks broke into fairly uniform grains under the action of sun exposure and temperature changes. The grain particles are removed from the parent rocks by wind or tremors, and they roll, slide, or saltate down, and scour the slope like a water flow, then the grain particles accumulate at the toe of the mountain and form a deposit fan.

In general a grain erosion site consists of three features: grain erosion surface at the top, grain flow section in the middle, and a deposit fan at the toe of the slope, as shown in Fig. 6.24 (Wang et al., 2009). The rock surface of grain erosion has a slope angle in the range of 45°–60°. There is no vegetation on the erosion surface.

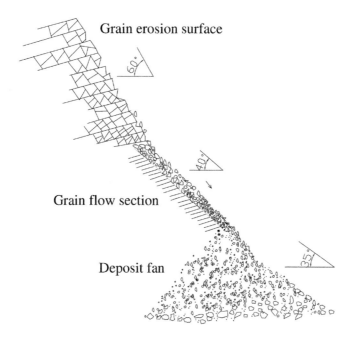

Grain erosion surface

Grain flow section

Deposit fan

Figure 6.21 Schematic diagram of a grain erosion site.

The detached particles roll or flow through a section, which has a slope angle of about 40°. The deposit fan has an angle of about 35°, which is equal to the angle of repose of the granular material.

In general a grain erosion site consists of three features: grain erosion surface at the top, grain flow section in the middle, and a deposit fan at the toe of the slope, as shown in Fig. 6.21 (Wang et al., 2009). The rock surface of grain erosion has a slope angle in the range of 45°–60°. There is no vegetation on the erosion surface. The detached particles roll or flow through a section, which has a slope angle of about 40°. The deposit fan has an angle of about 35°, which is equal to the angle of repose of the granular material.

In the grain flow section the particles roll and jump down the slope. At each step a particle hits against the slope material, thus, the grain flow scours the slope to form a flume like granular movement "path". The "path" has an angle of about 40° and the channel bed consists of relatively uniform sediment. Occasionally, particles fall onto the grain flow section from the grain erosion surface and initiate numerous particles rolling and saltating along the "path". Sometimes a layer of grains flows down the section, during which most of the particles slide and roll.

Figure 6.22(a) shows a grain erosion site in the Jinsha River Valley, which forms a 0.4 million m³ deposit fan. Figure 6.22(b) shows several grain erosion sites along the Minjiang River near Wenchuan. Avalanches induced by the Wenchuan Earthquake have left many bare rocks along the Minjiang River. Intensive grain erosion has been occurring, especially in the dry season from March to June in 2009. The grains are much finer than the avalanche deposits. A part of the grain erosion deposit has been

Figure 6.22 (a) Grain erosion in the Jinsha River Valley forming a 0.4 million m³ deposit fan; (b) Several grain erosion sites along the Minjiang River near Wenchuan; (c) Grain erosion on the north bank of the upper Jiangjia Ravine, in Yunnan province, China; (d) Grain flow on the slope of Jiuzhai Creek, which has killed many trees.

carried away by the river flow. Figure 6.22(c) shows a grain erosion site on the north bank of the upper Jiangjia Ravine, which provided a lot of loose solid material for debris flows. Figure 6.22(d) shows the vegetation damaged by grain erosion and grain flow along Jiuzhai Creek, which is a famous tourist attraction because of its beautiful landscape and vegetation. The grain flow section is 800 m long. The lithology consists of limestone and the grains generated from the erosion have a mean diameter of about 10 cm. The grains jump down the slope and hit against the trees. Most trees on the grain flow path have been killed.

Grain erosion has caused flying stones that have injured people. Most of the highways in western Sichuan province are constructed along rivers. People have repaired or reconstructed highways that were damaged or destroyed by the earthquake. All highways were reopened before the one year anniversary of the Wenchuan Earthquake.

(a)   (b)

Figure 6.23   (a) A slope debris flow from a grain erosion site by the Mianyuan River; (b) Grain erosion deposit causing local sedimentation in the Minjiang River.

Nevertheless, due to the continual grain erosion, particles with a diameter from 1 cm to 20 cm roll and saltate down the slope potentially falling on cars and people. The authors of this book have witnessed that the windshield of a car was broken and the driver was seriously injured by a flying stone. Grain erosion has caused many highways to become so called "flying stone section", especially the highway along the Minjiang River. The highway managers have hired many people to monitor the flying stones and issue warning signals. The highways are occasionally closed because of these flying stones.

Grain erosion provides plenty of solid materials for mass movements. The deposit fans consist of uniform and loose solid materials and have high slope. Rainfall with an intensity of more than 20 mm/day triggers mass movements of the grains. These mass movements behave like debris flows but the distance of movement is, however, much shorter than normal debris flows, and, in general, the travel distance is only several tens to one hundred meters. With water in the interstices of particles, which plays the role of lubrication, the grains move down the slope to streams or highways. Such a mass movement is called a slope debris flow. The slope debris flow carries a lot of grains into rivers or deposits the grains on highways, causing blockage of highway transportation or local sedimentation on the riverbed. Figure 6.23(a) shows a slope debris flow from a grain erosion site by the Mianyuan River. A rainfall of intensity of 47 mm/day initiated the slope debris flow, which carried sediment from the grain erosion deposit fan for a short distance from the high slope to the highway. The highway transportation was cut off. The angle of the debris flow deposit was about 10 degrees, much higher than a normal debris flow deposit. A lot of grain erosion has occurred along the Minjiang River. Figure 6.23(b) shows that a grain erosion deposit caused local sedimentation in the Minjiang River and changed the flow regime. The aquatic ecosystem was impacted by the sedimentation. Fish and benthic invertebrates lost their habitat due to the sedimentation.

(a)   (b)

Figure 6.24 (a) Bare limestone is cracked and broken down due to sun exposure and temperature changes; (b) A layer of a grain erosion deposit covering on an avalanche deposit fan on the Minjiang River near Wenchuan.

6.3.2 Mechanisms of grain erosion

Grain erosion occurs as a result of rock expansion and contraction due to temperature changes and breakdown of rocks under the action of sunshine. The new bare rocks are very vulnerable to erosion. Without the protection of vegetation cover or a layer of soil on the surface, bare rocks are acted on by the radiation from the sun and temperature changes. The exposure to weathering and the cycle of expansion during the day and contraction in the night causes fissures and breaks down the rocks. Figure 6.24(a) shows cracking bare limestone due to exposure to weathering and temperature changes along the Mianyuan River. The limestone is fragile and a surface layer about 10 cm thick was broken. The surface layer of the rock was further broken down into grains. Wind or tremors caused the grains to roll down the slope. Figure 6.24(b) shows a layer of a grain erosion deposit covering an avalanche deposit fan along the Minjiang River near Wenchuan. The grain erosion occurred on granite rock and the grains are generally finer than the grains in limestone areas. The grains are very uniform in size with a median diameter of about 1 cm. As a comparison the avalanche deposit beneath the grain layer is much more non-uniform consisting of stones of several meters and fine particles less than 1 mm. Because the grains from grain erosion are uniform in size and regular in shape people have mined the grains for building materials at some grain erosion sites with access to transportation facilities.

Grain erosion has occurred mainly in granite, limestone, and metamorphic rocks. The grains produced due to grain erosion in the limestone area were relatively coarse with diameters between 10–200 mm, grains in the granite area were finer with diameters between 5–30 mm; while in the metamorphic rock area the grain diameter varied in a large range from 0.1–300 mm. Figure 6.25 shows the size distributions of grain erosion deposits from the Minjiang River (granite rock), Jiuzhai Creek (limestone),

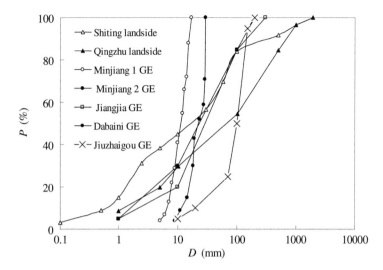

Figure 6.25 Size distributions of grain erosion deposits (GE represents grain erosion).

and Xiaojiang River (phyllite rocks), in which GE represents grain erosion. Minjiang 1 and Minjiang 2 are two grain erosion sites along the Minjiang River near Wenchuan. Jiangjia and Dabaini Ravines are two tributaries of the Xiaojiang River. As a comparison the size distributions of landslides on the Shiting and Qingzhu rivers, which are not far from Mianyang, are shown in the figure as well. The two landslides were triggered by the Wenchuan Earthquake and caused thousands of casualties. The range of the size distributions of solid particles in the landslide deposits has 6 orders of magnitude from 0.01 mm to 10 m, whereas the grain erosion deposits have diameters within 2 orders of magnitude for granite and limestone and 3 orders of magnitude for metamorphic rocks. Relatively uniform grain size and lack of large stones are common features of grain erosion which is much different from the grain size distribution of landslides and avalanches.

Exposure to weathering is the main agent for grain erosion, which may be proven by the following phenomena. The rocks along the Minjiang River became bare after the avalanches on May 12, 2008 triggered by the earthquake. Until early March 2009, grain erosion along the Minjiang River occurred only at several sites with limited areas. From March to June, the Minjiang River experienced the strongest sun exposure and driest season. During this time grain erosion developed very quickly with the area of grain erosion almost doubling. In general, the grain erosion on the south-facing bank of rivers is much more intense than that on the north-facing bank. Figure 6.26 shows the grain erosion on the south-facing bank and rill erosion (water erosion) on the north-facing bank of the Chaqing Gully along the Xiaojiang River. The lithology (phyllite) and rainfall on the two banks are the same but the sun exposureon the north-facing bank is much weaker than on the south-facing bank, therefore, grain erosion occurs on the south-facing bank and water erosion occurs on the north-facing bank. These phenomena prove that sun exposure weathering is the most important agent for grain erosion.

(a)   (b)

Figure 6.26 (a) Grain erosion occurs on the south-facing bank while (b) rill erosion (water erosion) occurs on the north-facing bank of a small stream.

As previously discussed in grain erosion wind detaches the grains from the rock and triggers grain flow. Thus, the amount and size of grains removed by wind is a function of wind speed. An experiment was done to study the relation of the amount and size of grains blown down with wind speed. Because grain erosion occurs at quite high elevations or on dangerous cliffs in the earthquake area, the experiment was done in the Xiaojiang River basin where grain erosion occurs on relatively small mountains. The lithology at the experimental sites is metamorphic rock consisting mainly of phyllite. The grains of a surface layer of rocks were blown down with bellows and batteries, which were transported to the mountain slopes along the Chaqing Gully by donkeys. The wind speed was measured with a rotational wind velocity meter. The bellows had a square nozzle of $10 \times 10 \, \text{cm}^2$. A blast of wind with a maximum speed of 20 m/s acted on the bare rock surface and the grains blown by the wind were collected with a bag and weighed with a balance. For each experiment an area of $1 \, \text{m}^2$ of the bare rock surface was acted on by the wind at a given wind speed for 10 min. Figure 6.27 shows the experimental results.

As shown in Fig. 6.27(a) the amount of grains blown down by wind in the four experiments had a consistent relation with wind speed and was proportional to the fourth power of wind speed. The size of the largest grain blown down from the bare rock was proportional to the wind speed (Fig. 6.27(b)), that is:

$$E_b = 0.00625 V^4; \quad D_m = V \tag{6.5}$$

in which V is the wind speed in m/s, E_b is the amount of the grains blown down by wind per time per square meter of bare rocks in g/min·m²; and D_m is the diameter of the largest grain blown down by wind. A wind with a speed of 20 m/s blew 1 kg of grains per minute away from one square meter of the bare rock with a maximum grain size of 20 mm. The bellows used in the experiment can only generate winds of

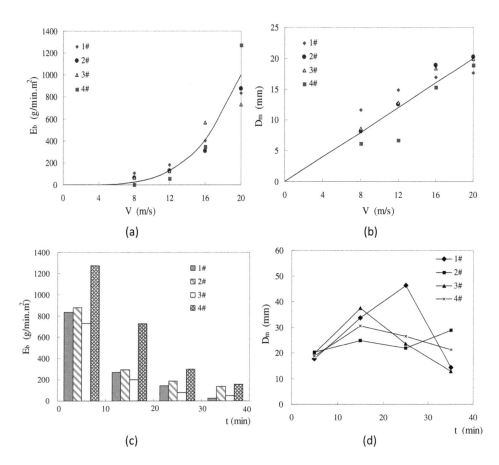

Figure 6.27 (a) Amount of grains blown by wind from 1 m² of rock surface per time as a function of the wind speed; (b) The largest grain size blown by wind as a function of wind speed; (c) Reduction with time of the amount of grains blown down by wind from 1 m² of rock surface per time in consecutive experiments; (d) Variation of the largest size of eolian grains in consecutive experiments (Wang et al., 2011).

speeds less than or equal to 20 m/s. In the gully, however, natural winds of maximum speeds over 35 m/s were measured with the wind velocity meter. The amount and size of grains blown down by wind can certainly be larger than those from the experiments.

The amount of grains blown down by wind depends on the wind speed and also on the cumulative time of weathering before the wind. In the experiment a wind speed of 20 m/s continued for 40 minutes, and the amount of grains blown down by the wind per time reduced with time, as shown in Fig. 6.27(c). Four measurements, each lasting 10 minutes, were made during the experiment of wind acting on 1 m² for 40 minutes. At the same wind speed the amount of grains blown down by wind per time reduced from 1000 g/m²min in the first 10 minutes to only 100 g/m² min in the fourth 10 minutes because after the grains on the top surface were removed the remaining grains were not readily detached from the rock. No general laws for the size of largest grains blown down by wind were found. It seems that the grain size increased from 20 mm to 45 mm, and, then, it reduced to about 25 mm as shown in Fig. 6.27(d).

The rate of grain erosion blown by wind per area per time in Fig. 6.27 represents a high instantaneous rate of grain erosion. The annual rate of grain erosion, however, depends on the area of fresh bare rocks, frequency of high speed winds, rate of sun weathering, and the action of temperature changes. In general, the bare rock may be eroded by several to several tens of centimeters per year, depending on the lithology, location, local weather, and winds. As shown in Fig. 6.24(b) the depth of the grain erosion deposit on the avalanche deposit fan, which could be easily identified by its uniform size, was measured with a scale at two or three places. The average depth of the grain erosion deposit multiplied by the surface area of the fan was the volume of grains eroded from each grain erosion site in the past year. The area of grain erosion of the bare rock surface was measured with laser range meters, which have a maximum error of 1m. The bare rock surfaces were generally larger than 100 m in length and width, therefore, the maximum relative error was less than 1%. The rate of grain erosion of rocks was obtained by dividing the volume of grains over the surface area of bare rock.

The same measurement was also performed for 9 grain erosion sites in the Xiaojiang River basin. The rate of grain erosion of bare rocks along on the Minjiang River was between 3 to 53 cm/yr, with an average rate of grain erosion of about 17 cm/yr. The rate of grain erosion in the Xiaojiang River basin was only 1.1–4.6 cm/yr, with an average rate of about 2.8 cm/yr (Wang et al., 2011). The rate of grain erosion in the earthquake area was much higher than in the Xiaojiang River basin because the bare rocks in the earthquake area were fresh. The rate of grain erosion will gradually reduce even if no control strategies are taken. Compared with shattering erosion the rate of grain erosion was more than 1000 times higher, but the rate will reduce quickly.

6.3.3 Control strategies

Studies have been devoted to the particle movement in the grain flow section and strategies have been suggested to control the grain flows (Wang et al., 2007a,b,c; Xu et al., 2007). The proposed control strategies were engineering measures for protection of highways, such as concrete sealing, protection walls, and removing grain deposits with machines (Xu et al., 2007). These strategies are aimed at controlling the movement of grains rather than controlling the erosion on the bare rocks. Nevertheless, grain flow control is not essential for mountain hazard control. If the grain erosion is not controlled, any control structure will finally fail to control grain flow. Thus, essentially no control has been achieved. Moreover, some engineering measures worsened the damage to highways (Sun et al., 2006).

The authors found from field investigations and experiments that if a thin layer of lichen and moss grows on the rock surface, the sun exposure and temperature changes are mitigated and cannot directly act on the rock, consequently no grain erosion occurs. Figure 6.28 shows the rock surface in the Chaqing Gully in the Jinsha River basin, in which the left part was suffering grain erosion, while on the right part moss and lichens have been established. An experiment was done on the bare rocks. As the wind acted on the left part at a speed of 20 m/s from the experimental blowers, more than 9 kg of grains were blown down in 10 minutes from 1 m^2 of rock surface. When the wind acted on the right part, however, no grains were blown away because of the protection

Figure 6.28 A layer of moss and lichens protects the rock (phyllite) from grain erosion.

of the moss and lichens. The moss and lichen layer was only about 1 mm thick, but it effectively protected the rock from direct action of sun exposure and temperature changes, therefore, the rocks were not broken down.

The essential cause of grain erosion is devegetation and exposure of bare rocks to sun weathering and temperature changes. Therefore, an essential strategy to control erosion is revegetation. Several studies have been done on the interactions between moisture, lithobiontic organisms, and rock weathering. Some researchers simply assumed that organic weathering replaces inorganic processes and paid attention to the rates of bioerosion while little attention was paid to the role of erosion control played by a particular vegetation species (Naylor et al., 2002). Only a few studies have attempted to identify bioprotection provided by lichens during weathering processes (Carter and Viles, 2003, 2005). Yet there is little consensus over whether rates of lichen-mediated weathering are slower than rates of abiotic weathering of otherwise identical rock surfaces (Lee, 2000). In fact, epilithic organisms can tremendously change microclimate. The canopy temperature of cushion vegetation in the Alps could be 27°C and relative humidity could be 98%, while the air temperature is 4°C and relative humidity is only 40% (Korner 2003). Scientists also conducted field experiments and concluded that epilithic lichens effectively retain moisture and reduce thermal stress on the surface of limestone (Carter and Viles, 2003).

An experiment was done at Xiaomuling, which is a grain erosion site along the Mianyuan River. Spores of five moss species were collected from local and neighboring areas and mixed with a clay suspension. The clay material was collected from fine sediment deposits on the floodplains of the Mianyuan River and the clay particles were finer than 0.01 mm. The concentration of clay suspension was 265 kg/m^3. The

(a)   (b)   (c)

Figure 6.29   (a) Grain erosion site at Xiaomuling on the Mianyuan River, Sichuan province; (b) Clay
suspension with moss spores was splashed on the Xiaomuling rock subject to grain erosion;
(c) Two months later the rock surface was green and grain erosion was controlled.

collected sporophyls were smashed with a machine. The clay suspension had a certain concentration of the smashed sporophyls to have a sufficient amount of spores per liter of clay suspension. The experimental plots were more than 100 m above the Mianyuan River. Local farmers were hired to carry the clay suspension with moss spores up to the mountain and pour down the suspension onto the bare rock surface at several plots.

Several rainstorms occurred after the experiment. Rain washed a part of the clay suspension down to the lower part of the bare rock, which helped to spread the moss spores to a large area of the rock surface. Figure 6.29(a) shows the grain erosion site at Xiaomuling on the Mianyuan River. Figure 6.29(b) shows the clay suspension layer with moss spores. Two moss species had germinated on the rock surface after one month of the experiment. Two months later the experimental plots had become green with a thin layer of moss growing, as shown in Fig. 6.29(c). Because rainstorms spread the clay suspension to the lower part of the rocks, the area of greened rock surface was larger than the rock surface with clay cover at the beginning of the experiment. Only two species of moss: *Rhizomnium sp.* and *Grimmia sp.* successfully germinated on the rock surface. Both of the species were collected from local vegetated rocks.

Selection of moss species is important for the success of vegetation restoration. However, the clay suspension with moss spores must be poured onto the top part of the local bare rock. If only a lower part of the bare rock is covered with clay suspension and it is revegetated, while the grain erosion on the upper part continues, the grains falling from upper part may destroy or bury the newly greened lower rock surface. The grain erosion rock surfaces in the Minjiang River valley are rather high and very dangerous for humans to climb up. Helicopters may be used to pour the clay suspension with moss spores onto the top of the bare mountains in order to quickly control grain erosion and revegetate the bare mountains.

REFERENCES

Burbank, D.W., Leland, J., Fielding, E., Anderson, R.J., Brozovic, N., Reid, M.R., and Duncan, C. 1996. Bedrock incision, rock uplift and threshold hillslopes in the northwestern Himalayas. Nature, 379, 505–510.

Carter, N.E.A., and Viles, H.A. 2003. Experimental investigations into the interactions between moisture, rock surface temperatures and an epilithic lichen cover in the bioprotection of limestone. Building and Environment. 38(9–10), 1225–1234.

Carter, N.E.A., and Viles, H.A. 2005. Bioprotection explored: The story of a little known earth surface process. Geomorphology. 67(3–4), 273–281.

Chang, X., Wu, G.X., and Guo, Q. 2006. Formation of sand-sliding slope and its controlling countermeasure Highway. 1, 89–91 (in Chinese).

Curry, A.M., and Black, R. 2002. Structure, sedimentology and evolution of rockfall talus in mynydd du, South Wales. Proceedings of the Geologists Association. 114, 49–64.

Finnegan, N.J., Hallet, B., Montgomery, D.R., Zeitler, P.K., Stone, J.O., Anders, A.M., and Liu, Y. 2008. Coupling of rock uplift and river incision in the Namche Barwa-Gyala Peri massif, Tibet. Geological Society of America Bulletin, 120(1–2), 142–155.

Garcia-Ruiz, J.M., Valero-Garces, B., Gonzalez-Samperiz, P., Lorente, A., Marti-Bono, C., Begueria, S., and Edwards, L. 2001. Stratified scree in the central Ppanish Pyrenees: Palaeoenvironmental implications. Permafrost and Periglacial Processes. 12(3), 233–242.

Garzanti, E., Vezzoli, G., Ando, S., France-Lanord, C., Singh, S.K., and Foster, G. 2004. Sand petrology and focused erosion in collision orogens: The Brahmaputra case. Earth and Planetary Science Letters, 220(1–2), 157–174.

Goldman, S.J., Jackson, K., and Bursztynsky, T.A. 1986. Erosion and Sediment Control Handbook. McGraw-Hill Book Company, New York.

Columbia University. 2000. Columbia Encyclopedia. Columbia University Press.

Halsey, D.P., Mitchell, D.J., and Dews, S.J. 1998. Influence of climatically induced cycles in physical weathering. Quarterly Journal of Engineering Geology and Hydrogeology. 31, 359–367.

Harris, S.A. and Prick, A. 2000. Conditions of formation of stratified screes, Slims River Valley, Yukon territory: A possible analogue with some deposits from Belgium. Earth Surface Processes and Landforms. 25(5), 463–481.

He, P., Guo, K., Gao, J.X., Shi, P.J., Zhang, Y.Z., and Zhuang, H.X. 2005. Vegetation types and their geographic distribution in the source area of the Yarlung Zangbo. Journal of Mountain Research. 23(3):267–273. (in Chinese)

Hetu, B., Vansteijin, H., and Bertran, P. 1995. Role of dry grain flows in the foemation of a certain type of stratified scree. Permafrost and Periglacial Processes. 6(2), 173–194.

Hou, F. and Wang, L. 2009. The wild orchidaceae resources in Motuo County and resource exploitation. Forest By-Product and Speciality in China, 102(5), 77–80. (in Chinese)

Journal of Engineering Geology and Hydrogeology, 31: 359–367.

Korner, C. 2003. Alpine Plant Life: Functional Plant Ecology of High Mountain Ecosystems. Springer Press, Berlin, Heidelberg, New York.

Lee, M.R. 2000. Weathering of rocks by lichens: Fragmentation, dissolution and precipitation of minerals in a microbial microcosm. Environmental Mineralogy: Microbial Interactions Anthropogenic Influences, Contaminated Land and Waste Management. 9, 77–107.

Lu, H. 2008. Research on land use/cover change and its environment impact in Naidong County, Tibet. Master's Thesis, Capital Normal University, Beijing. (in Chinese)

Matsuoka, N. 2008. Frost weathering and rockwall erosion in the southeastern Swiss Alps: Long-term (1994–2006) observations. Geomorphology. 99(1–4), 353–368.

Miehe, G., Miehe, S., Schlutz, F., Kaiser, K., and Duo, L. 2006. Palaeoecological and experimental evidence of former forests and woodlands in the treeless desert pastures of Southern Tibet

(Lhasa, AR Xizang, China). Palaeogeography, Palaeoclimatology, Palaeoecology, 242(1–2), 54–67.

Naylor, L.A., Viles, H.A., and Carter, N.E.A. 2002. Biogeomorphology revisited: Looking towards the future. Geomorphology. 47(1), 3–14.

Saas, O. and Krautblatter, M. 2007. Debris flow-dominated and rockfall-dominated talus slopes: Genetic models derived from gpr measurements. Geomorphology. 86(1–2), 176–192.

Shang, Y., Yue, Z., Yang Z., Wang, Y., and Liu D. 2003. Addressing severe slope failure hazards along Sichuan-Tibet Highway in Southwestern China. Episopdes, 26(2): 94–104.

Sun, H.Y., Shang, Y.Q., and Lu, Q. 2006. Formative mechanism and control countermeasure of sand-sliding slope. Journal of Natural Disasters. 15(4), 28–32.

Sun, H., Zhou, Z.K., and Yu, H.Y. 1997. The vegetation of the Big Bend Gorge of Yalu Tsangpo River, S. E. Tibet, E. Himalayas. Acta Botanica Yunnanica, 19(1), 57–66. (in Chinese)

Sun, M., Shen, W.S., Li, H.H., Zhang, H., and Sun, J. 2010. Traits and dynamic changes of the aeolian sandy land in the source region of the Yarlung Zangbo River in Tibet. Journal of Resources, 25(7), 1163–1171. (in Chinese).

Wang, C.H., Que, Y., Li, X.P., and Zhang, X.G. 2007. Movement characteristics and dynamical numerical analysis of sand-sliding slope composed by granular clastics: Part ii of sand-sliding slope series. Rock and Soil Mechanics. 28(2), 219–223 (in Chinese with English abstract).

Wang, X.D. and Wang, Z.Y. 1999. Effect of land use change on runoff and sediment yield. Journal of Sediment Research, 14(4), 37–44.

Wang, Z.Y., Cui, P., and Wang, R.Y. 2009. Mass movements triggered by the Wenchuan Earthquake and management strategies of quake lakes. International Journal of River Basin Management. 7(1), 1–12.

Wang, Z.Y., Cui, P., Yu, G.A., and Zhang, K. 2010. Stability of landslide dams and development of knickpoints. Environmental Earth Sciences, 65(4), 1067–1080.

Wang, Z.Y., Huang, G.H., Wang, G.Q., and Gao, J. 2004. Modeling of vegetation-erosion dynamics in watershed systems. Journal of Environment Engineering, 130(77), 792–800.

Wang, Z.Y., Shi, W.J., and Liu, D.D. 2011. Continual erosion of bare rocks after the Wenchuan Earthquake and control strategies. Journal of Asian Earth Sciences, 40(4), 915–925.

Wang, Z.Y., Wang, G.Q., and Huang, G.H. 2008. Modeling state of vegetation and soil erosion over large areas. International Journal of Sediment Research. 23, 181–196.

Wen, A.B., Liu, S.Z., Fan, J.R., Zhu, P.Y., Zhou, L., Zhang, X.B., Zhang, Y.Y., Xu, J.Y., and Bai, L.Z. 2000. Soil erosion rate using 137Cs technique in the middle Yarlungtsangpo. Journal of Soil and Water Conservation, 14(4), 47–50. (in Chinese).

Xu, J., Wang, C.H., He, S.M., Zhang, X.G., and Zhou, L. 2007. Sheet pile wall's techniques for stabilization of sand-sliding slope composed of granular clast. Research of Soil and Water Conservation. 14(3), 315–317.

Zhang, C.L., Zou, X.Y., Yang, P., Dong, X.Y., Li, S., Wei, X.H., Yang, S., and Pan, X.H. 2007. Wind tunnel test and 137Cs tracing study on wind erosion of several soils in Tibet. Soil and Tillage Research, 94(2), 269–282.

Zhang, J., Tan, J.L., Deng, L.L., Li, S.Z., and Huang, G.L. 2008. Investigation and assessment of the terrestrial vegetation in the middle reaches of the Yarlung Zangbo River in Tibet. Forest Resources Management, 4, 118–123.

Zhao, J. and Li, R. 2008. Soil erosion and subarea characteristics in Yarlung Tsangpo River basin. Journal of Yangtze River Scientific Research Institute, 25(3), 42–45. (in Chinese)

Zhao, W.Z., Zhang, Z.H., and Li, Q.Y. 2007. Growth and reproduction of Sophora moorcroftiana responding to altitude and sand burial in the middle Tibet. Environmental Geology, 53(1), 11–17. (in Chinese)

Zhao, Y.Z., Zou, X.Y., Cheng, H., Jia, H.K., Wu, G.Y., Zhang, C.L., and Gao, S.Y. 2006. Assessing the ecological security of the Tibetan plateau: Methodology and a case study for Lhaze County. Journal of Environmental Management, 80(2), 120–131.

Zhen, W.L. 1999. Analyses of the flotistic features on the families, genera of pteridophyta from the Big Bend region at Yalu Tsangpo (river), Xizang (Tibet), China. Acta Botanica Yunnanica, 21(1), 43–50. (in Chinese)

Zhong, C., He, Z.Y., and Liu, S.Z. 2005. Evaluation of eco-environmental stability based on GIS in Tibet, China. Wuhan University Journal of Natural Sciences, 10(4), 653–658.

Chapter 7

Aquatic ecology

7.1 AQUATIC ECOLOGY IN THE YARLUNG TSANGPO BASIN

Describing and understanding the patterns in biological diversity along major geographical gradients is very important in ecological research because geographical gradients affect aquatic assemblages (Jacobsen, 2004). Unfortunately, most previous studies of aquatic ecology and the impact of environmental parameters were conducted mainly in lowland rivers in China (Duan et al., 2011; Pan et al., 2011, 2012). The authors of this book conducted investigations of aquatic ecology in the Yarlung Tsangpo River located in the highest elevation area of the world. The Yarlung Tsangpo basin is bordered by the Himalaya Mountains in the south and the Gangdisê and Nyainqêntanglha Mountains in the north. It is 1200 km long and flows from west to east in the South Tibet Valley. In East Tibet, the river flows through the world's largest and deepest canyon, the Yarlung Tsangpo Grand Canyon. After leaving the canyon, the river flows into India, and named as the Brahmaputra River.

The aquatic ecology in the Yarlung Tsangpo River is unique compared to other typical mountain rivers located at elevations lower than 2000 m, due to the drastic changes in the environmental conditions from higher to lower elevations, such as climate, riparian vegetation, and water temperature (Füreder et al., 2001). As the elevation of the river descends from 4500 m to 3500 m and then to 2000 m, the riparian vegetation changes from cold desert to arid steppe and then to deciduous scrub, respectively. Temperature and precipitation within the Yarlung Tsangpo basin also vary greatly with the elevation. The annual precipitation in the upstream portion of the river basin is less than 300 mm, while in the downstream reach of the Yarlung Tsangpo Grand Canyon, the annual precipitation is 4000 mm (Liu et al., 2007). The average annual air temperature in the downstream reach is at least 10°C higher than that in the upstream reach (Song et al., 2011).

The Yarlung Tsangpo River and its tributaries have the most remote and undisturbed aquatic environment. They are much less impacted by pollutions from agriculture and sewage water than rivers in other parts of China. However, the Yarlung Tsangpo River and its tributaries are especially sensitive to climate change. According to monitoring data, during 1961–2005, the magnitude of climate warming in the Yarlung Tsangpo basin exceeded that of the Mt. Everest region, the Qinghai-Tibet Plateau, and the global average. The magnitude of precipitation increase and the potential for evapotranspiration decrease were also greater in the Yarlung Tsangpo basin. As a result, the source of the water in the Yarlung Tsangpo basin will change (You et al.,

2007). Since most of the water source for the rivers at high elevations is the glaciers, the aquatic biota in these rivers have adapted to persistently low temperatures (Skjelkvåle and Wright, 1998), and would react very sensitively to the change of water source. Therefore, it is critical to study the characteristics of the aquatic ecosystems in the river basins at high elevations.

From previous studies, researchers have gained increased knowledge about the characteristics of macroinvertebrate assemblages in rivers at different elevations. Jacobsen (2004) indicated that family-level identification of macroinvertebrates can facilitate interpretation of variations of biodiversity along geographic gradients. Studies on macroinvertebrate assemblages in remote highland lakes across Europe have shown that the response of the macroinvertebrates to gradients varied with elevation (Fjellheim et al., 2009).

The occurrence of macroinvertebrate families and species in glacier-fed rivers is determined by the maximum water temperature and channel stability (Milner et al., 2010). Reduced water temperatures from increased contributions of glacial meltwater and decreased channel stability from changing runoff patterns reduce the diversity of macroinvertebrate assemblages in glacier-melt dominated rivers (McGregor et al., 1995). A preliminary investigation of the macroinvertebrates in the Yarlung Tsangpo River showed that the density of the macroinvertebrates was much lower and the assemblage composition was quite different than that observed in rivers at lower elevations (Zhao and Liu, 2010). However, the distribution of the macroinvertebrate assemblages in the Yarlung Tsangpo basin at different elevations is yet to be studied.

7.1.1   Investigation of benthic macroinvertebrates

Field investigations and macroinvertebrate samplings were conducted from October 2009 to June 2010 in the Yarlung Tsangpo River basin. Figure 7.1 shows the entire Yarlung Tsangpo River basin and the sampling locations in the stem and its tributaries: Nianchu, Lhasa, Niyang, and Parlong Tsangpo Rivers. The sites S6, S9, S11, S12, and S13 were located in the main channel of the Yarlung Tsangpo River. Site S3 was in the upstream tributary, the Nianchu River. Sites S1, S4, and S8 were in the midstream tributary, the Lhasa River. Sites S2, S10, and S14 were in the mid-downstream tributary, the Niyang River. Sites S5 and S7 were in the downstream tributary, the Parlong Tsangpo River.

Three replicate samples of macroinvertebrates were collected at each sampling site. Each sample was collected at a different place with a sampling area of $1/3\,m^2$. Macroinvertebrates were collected using a kick-net with mesh of $420\,\mu m$ in diameter. Specimens were manually sorted from sediment on a white porcelain plate and preserved in 75% ethanol. Wet weight of the animals was determined with an electronic balance after the animals were blotted. Then the dry weight (mollusks without shells) was calculated according to the ratios of dry-wet weight and tissue-shell weight reported by Yan and Liang (1999). All taxa were assigned to functional feeding groups (shredders, collector-gatherers, collector-filterers, scrapers, and predators) (see Morse et al., 1994; Liang and Wang, 1999). When a taxon had several possible feeding activities, its functional designations were equally proportioned (i.e., if a taxon can be both collector-gatherer and scraper, its abundance was divided 50%:50% into these groups).

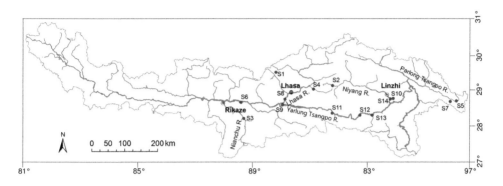

Figure 7.1 Study area and sampling sites in the Yarlung Tsangpo basin.

Physic-environmental parameters were measured at the sampling sites. Water depth (Z) and Secchi depth (Z_{SD}) were measured with a sounding lead and a Secchi Disc, respectively. Water velocity (U) was measured with a propeller-type current meter (Model LS 1206B). Water samples taken near the surface and at the bottom were combined for laboratory analyses. Suspended sediments (SS) were analyzed following procedures outlined by American Public Health Association (APHA), 2002. Conductivity was measured using a conductivity meter (Model DDS-11A). Total nitrogen (TN) was analyzed using the alkaline potassium persulfate digestion-UV spectrophotometric method. Total phosphorus (TP) was analyzed using the ammonium molybdate method. All parameters were analyzed according to Standard Methods for Water and Wastewater Monitoring and Analysis (APHA, 2002).

Taxa richness, S (the number of species), is the most important characteristic of biodiversity as it provides a measure of both ecological diversity and habitat conditions of streams. The Shannon-Weaver index, H', integrates the taxa richness and evenness of the distribution of animals for different species. It is defined by Krebs (1978) as:

$$H' = -\sum_{i=1}^{S} \frac{n_i}{N} \ln\left(\frac{n_i}{N}\right) \tag{7.1}$$

where S is the number of species (richness), N is the total number of individual animals, and n_i is the number of individual animal of the ith species. The Shannon-Wiener bio-index reflects the taxon richness and the evenness of the number distribution of species. However, the Shannon-Wiener Index provides no information on the total abundance of the bio-community, and sometimes the index even provides incorrect results. The numbers of individual animals per unit area vary substantially for habitats with different physical conditions. Considering both abundance and biodiversity, a biocommunity index, B, was used (Wang et al., 2008):

$$B = -\ln N \sum_{i=1}^{S} \frac{n_i}{N} \ln\left(\frac{n_i}{N}\right) \tag{7.2}$$

In addition, the beta diversity β was evaluated for the assessment of the regional diversity of the entire Yarlung Tsangpo River basin. It is defined as:

$$\beta = \frac{M}{\frac{1}{S} \sum_{i=1}^{S} m_i}, \tag{7.3}$$

where M is the number of the sampling sites with the highest variation in macroinvertebrate compositions. m_i is the number of sites where the ith taxon exists (Wang et al., 2012). For a given M, the higher the value of β is, the higher the heterogeneity of the community is.

The taxa richness S and the improved Shannon-Wiener index B have been considered to be the most important indices for evaluating the alpha diversity for local sampling sites, however, density D and biomass W also have been widely used for evaluations of the alpha diversity (Duan et al., 2009). Density D is the total number of macroinvertebrate individuals per unit area, in ind./m^2. Biomass W is the total wet mass of macroinvertebrates per unit area, in g/m^2. Density D and biomass W were chosen by the authors of this book to assess the aquatic ecology and analyze the impact of elevation gradients on the biodiversity.

Macroinvertebrates were also sampled from the Juma River, which is a mountain river basin in the suburb of Beijing with an elevation less than 300 m. The sampling sites in the Juma basin shared similar substrate compositions and flow conditions as those in the Yarlung Tsangpo River basin. Alpha and beta diversity indices of the macroinvertebrate assemblages in both basins were compared. For calculation of the beta diversity index, the eight sites which had the highest variation in macroinvertebrate compositions were selected, giving $M = 8$.

The cumulative taxa richness was calculated to analyze the impact of elevation on the regional taxa distribution. Cumulative taxa richness S_n was introduced and calculated as follows: S_1 was the taxa richness at site $S1$, and S_2 was the total taxa richness in the samples of $S1$ and $S2$. S_n was the total taxa richness in all of the samples of $S1$, $S2$, . . . , and S_n. As listed in Table 7.1, $S1$ had the highest elevation, $S2$ had the second highest elevation, and Sn had the n th highest elevation. The taxa that occurred at $S2$ but not at $S1$ were tallied as dS_2, thus $S_2 = S_1 + dS_2$. Similarly, the taxa that occurred at S_n but not at $S1$, $S2$, . . ., $S_{(n-1)}$ were tallied as dS_n, then $S_n = S_{(n-1)} + dS_n$. In this way, the cumulative loss and gain of macroinvertebrate families could be calculated as the elevation decreased.

In addition to the analysis of the impact of elevation on the macroinvertebrate assemblages, taxonomic distribution along the gradients of the other main environmental parameters was also determined using a method of indirect gradient analysis, Detrended Correspondence Analysis (DCA) (Hill, 1979). The software CANOCO for Windows Package 4.5 (Ter Braak and Šmilauer, 2002) was used for the DCA. Clusters of samples in an ordinate space were identified by giving a relatively objective community ordination of the matrix of [species, samples].

7.1.2 Environmental parameters

Table 7.1 lists the environmental parameters, including the elevation (E), water depth (h), flow velocity (v), dissolved oxygen concentration (DO), water temperature (T),

Table 7.1 Environmental parameters and macroinvertebrate biodiversity indices at the sampling sites.

Site	E/m	h/m	v/(m·s⁻¹)	DO/(mg·L⁻¹)	T/°C	Season	Stream condition	Riparian condition	S	B	DI/(ind·m⁻²)	WI/(g·m⁻²)
S1	4484	0–0.2	0.3–0.5	8.4	1.3	Winter	Step-pool developed, channel width = 3 m	Alpine meadow, VC = 100%, VH = 1–5 cm	21	13.4	620	2.47
S2	4228	0.1–0.4	0.83	8.1	3.5	Winter	Stable bed, channel width = 10 m	Alpine meadow, VC = 100%, VH = 1–5 cm	21	14.5	672	6.12
S3	4014	0–0.25	0.3–0.5	7.9	4.7	Winter	Braided stream, macro-algae covered bed, channel width = 100 m	Channelized bank, VC <5%	17	10.1	375	0.27
S4	3916	0–0.15	0.1–0.3	6.2	10.5	Winter	Wetland linked with channel, rich humus, channel width = 50 m	Alpine meadow, VC = 100%, VH = 5–20 cm	17	10.5	186	1.17
S5	3901	0.2–0.4	0.1	7.9	3.0	Summer	Barrier lake, sand bed, channel width = 5 m	No vegetation	8	3.1	192	0.26
S6	3768	0–0.3	0.3	9.6	17.4	Winter	Braided stream, gravel bed, channel width = 50 m	No vegetation	29	14.1	346	2.72
S7	3752	0.1–0.3	0.3–0.5	8.5	2.0	Summer	Glacier-fed stream, gravel bed, channel width = 10 m	Shrub, VC = 10%	36	15.7	326	1.25
S8	3598	0–0.5	0.3–0.8	10.0	10.5	Winter	Braided river, wide valley, gravel bed, channel width = 50 m	Herbage and trees, VC = 100%	25	17.4	830	21.35
S9	3566	0–0.5	0.3	7.7	12.4	Winter	Braided river, wide valley, gravel-clay bed, channel width = 40 m	No vegetation	17	12.2	2440	7.59
S10	3514	0–0.4	0.3–0.5	8.1	7.2	Winter	Step-pool developed, channel width = 5 m	Channelized bank, VC <2%	33	20.4	1513	3.33
S11	3237	0.5–1.5	0.5–1.5	8.0	8.0	Summer	Step-pool developed, channel width = 20 m	No vegetation	18	13.1	279	0.51
S12	2993	0.1–0.5	0.0	6.7	13.0	Summer	River bend, lentic, channel width = 100 m	No vegetation	14	9.5	680	2.12
S13	2959	0.2–0.4	1.5–2	7.8	8.0	Summer	Gravel bed, lotic, channel width = 200 m	Alpine meadow, VC = 50%, VH = 5–200 cm	16	7.2	46	0.24
S14	2948	0.3–1.0	0.1–0.3, 1.5–2.0	7.5	15.0	Summer	Wetland, gravel bed, channel width = 30 m	Alpine meadow, VC = 80%, VH = 1–10 cm	20	10.4	2415	2.64

Note: E: elevation; h: water depth; v: flow velocity; DO: dissolved oxygen; T: water temperature; VC: riparian vegetation coverage rate; VH: riparian vegetation height; S: taxa richness; B: improved Shannon–wiener index; D: macroinvertebrate individual density; and W: biomass.

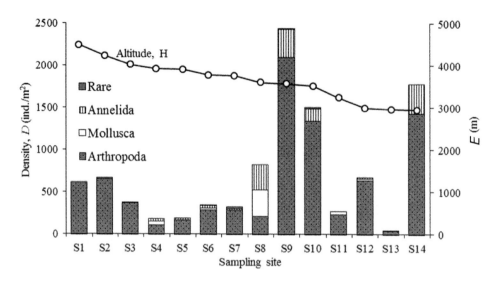

Figure 7.2 Density compositions of different groups of the macroinvertebrates at the sampling sites.

and the coverage rate (VC) and height (VH) of riparian vegetation for all sampling sites. The alpha diversity indices S, B, D, and W are also included in Table 7.1.

7.1.3 Composition of macroinvertebrate assemblages

A total of 110 macroinvertebrate taxa belonging to 57 families and 102 genera were identified in the Yarlung Tsangpo River basin. Among these macroinvertebrates, there were 1 Turbellaria, 1 Nematoda, 16 Oligochaeta, 4 Hirudinea, 7 mollusks, 1 Arachnida, 1 Crustacea, and 79 Insecta. The total taxa richness was much higher than that previously reported by Zhao and Liu (2010). The proportions of Oligochaeta, Hirudinea, Gastropoda, and Insecta were 14.5%, 3.6%, 5.5%, and 70%, respectively. Figure 7.2 shows the density compositions of the taxonomic groups Arthropoda, Mollusca, Annelida, and other rare groups at the sites. Arthropoda was the most dominant group in the entire basin. Mollusca and Annelida were mainly at the sites at elevations lower than 3700 m. The proportions of Annelida or Mollusca were extremely low at the sites at elevations higher than 4000 m. The rare groups including Platyhelminthes and Nematoda were only found at the sites at elevations higher than 3500 m.

Figure 7.3 shows the density compositions of the five functional feeding groups: shredders, scrapers, collector-filterers, collector-gatherers, and predators. Collector-gatherers and predators inhabited at almost all sites. Scrapers mainly inhabited at the sites at elevations 3500–4500 m. Shredders mainly inhabited at the sites at elevations 2900–3600 m and 4000–4500 m. Trees and bushes grew at elevations 2900–3600 m, adequately supplying fallen leaves for the shredders, i.e., Chrysomelidae. In addition, alpine algae and plateau meadows grew at elevations 4000–4500 m, sufficiently

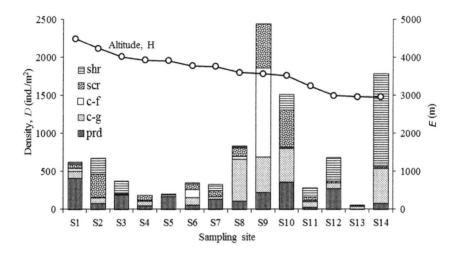

Figure 7.3 Density compositions of different functional feeding groups of macroinvertebrates.

supplying food for the shredders, i.e., *Tipula sp.*, Pteronarcidae. Collector-filterers inhabited at the sites at elevations 3500–3900 m. Predators existed at all sites, with an increased density proportion as the elevation increased. Tomanova et al. (2007) concluded that the density proportions of collector-gatherers, shredders, and scrapers were clearly related to the elevation. Macroinvertebrates are constantly moving, and the distance that they cover depends on the sparsity of their food source distribution (Boyero, 2005). The food sources for the shredders, scrapers, and collectors in the Yarlung Tsangpo River basin rely on the vegetation distribution, which is affected by the elevation; thus, the distributions of these three groups vary with the elevation. On the other hand, the primary food source for predators are living animals. And, therefore, predators are distributed at all the elevations with living animals.

As listed in Table 7.1, the highest taxa richness and improved Shannon-Wiener index were at S7 (36 and 15.7, respectively) and S10 (33 and 20.4, respectively), and the lowest taxa richness and improved Shannon-Wiener index were at S5 (8 and 3.1, respectively). The taxa richness in the upstream tributary, the Nianchu River, was lower than that in the main channel of the Yarlung Tsangpo River. The average taxa richness in the mid- and downstream tributaries, the Lhasa, Niyang, and Parlong Tsangpo Rivers, were higher than that in the main stem Yarlung Tsangpo River. The two highest densities of the macroinvertebrates were 2440 ind./m² at S9 and 2415 ind./m² at S14, respectively. The three lowest densities were 46 ind./m² at S13, 186 ind./m² at S4, and 192 ind./m² at S5, respectively.

Figure 7.4 compares the alpha and beta diversity indices of the macroinvertebrate assemblages in the Yarlung Tsangpo and Juma basins. Both had similar alpha diversity indices; however, the beta diversity was much higher in the Yarlung Tsangpo River than that in the Juma River. This indicates that the heterogeneity of the macroinvertebrate assemblages and the regional diversity of aquatic ecosystems are much higher in highland rivers than those in lowland rivers.

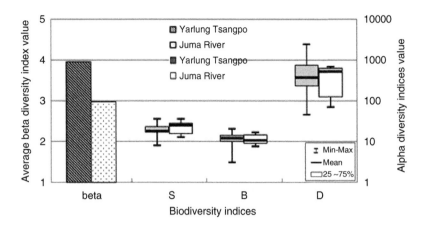

Figure 7.4 Comparison of biodiversity indices of macroinvertebrate assemblages in the Yarlung Tsangpo and Juma basins.

7.1.4   Effect of elevation gradients on biodiversity of macroinvertebrates

Elevation is considered to be the most important variable for determining the living conditions of macroinvertebrates in plateau areas (Čiamporová-Zaťžovčová et al., 2010). In Europe, it was found that the taxa richness of macroinvertebrates generally decreased as the elevation increased (Brittain and Milner, 2001). However, it was reported that there was an increased number of macroinvertebrate families in the rivers at high elevations in South America (Henriques-Oliveira and Nessimian, 2010). The authors of this book have found out that the alpha diversity indices are clearly related to the elevation.

Figure 7.5(a) shows the taxa richness S as a function of the elevation E. S increased as E increased in the range of 2900–3500 m, while S decreased and then stayed at low values in the range of 3700–4900 m. Figure 7.5(b) shows the improved Shannon-Wiener index B as a function of the elevation E, which is similar to the relation between S and E. The highest value of B (combines the information of both density and taxa richness) occurred at approximately elevation 3500 m, indicating that both density and taxa richness of the macroinvertebrate assemblages reached their highest values at these elevations.

Figure 7.6 shows the density D and biomass W as functions of the elevation E. Similar to the relation between B and E, the density and biomass fluctuated with the elevation: firstly, D and W increased as the elevation increased from 2900 to 3500 m; afterwards, D and W decreased as the elevation increased from 3500 to 4000 m; and then D and W increased as the elevation increased from 4000 to 4500 m. The high density and biomass at the high elevations were due to the extreme dominance of the few high elevation tolerant taxa, such as *Physa Draparnaud*, *Hippeutis* sp., *Rhyacodrilus stephesoni*, Hydrachnidae, Baetidae, Ecdyuridae, and Glossosomatidae. The maximum values of D and W also occurred around 3500 to 3700 m. The values

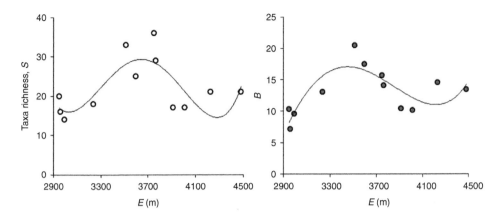

Figure 7.5 (a) Taxa richness S as a function of Elevation E; (b) Index B as a function of Elevation E.

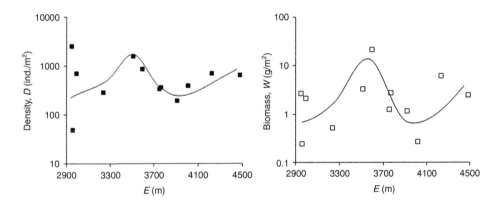

Figure 7.6 (a) Density D as a function of the elevation E; (b) Biomass W as a function of the elevation E.

of D and W varied significantly when the elevation was between 2900 and 3300 m. The possible reason for the high variance was that the extremely high flow velocity at the mid-downstream sites restrained the density of the macroinvertebrate assemblages.

Figure 7.7 shows the cumulative taxa richness S_n as a function of the elevation E. The cumulative taxa richness decreased as the elevation increased. Similar relations were found in northern Ecuador, South America (Jacobsen, 2004), where S_n increased slowly as the elevation decreased from 4500 to 4000 m, and then increased quickly as the elevation decreased from 4000 to 3500 m. The richness increased slowly as the elevation decreased from 3500 to 2900 m. In the Yarlung Tsangpo basin, the area at elevation 3500 to 4000 m happens to be where the vegetation belt varies from highland steppe to complex vegetation with grasses, herbs, shrubs, and woods. In addition, the main water source for the rivers in regions at elevations higher than 4000 m comes

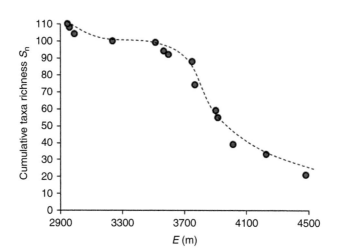

Figure 7.7 Relation between cumulative taxa richness S_n and elevation E.

from melting snow and ice, whereas the main water source for rivers at elevations lower than 3500 m is rainfall (Füreder et al., 2001). Milner et al. (2010) stated that there is a large variance in macroinvertebrate assemblages found in rivers with different water sources. For example, the winter stonefly Capniidae was only found in rivers at elevations ranging from 3500 to 3700 m. Lu et al. (2011) pointed out that the river patterns were different for rivers at elevations below and above 3500 m in the Hindu Kush-Himalayas area. Therefore, the sharp slope at elevations 3500 to 4000 m shown in Fig. 7.7 is believed to be caused by the variation of food resource availability and the changes in the geomorphology and water resources at different elevations.

The authors analyzed the relation between the cumulative loss and gain of families in the Yarlung Tsangpo River and the elevation. As shown in Fig. 7.8, the cumulative loss of families increased linearly from around 10 to 50 as the elevation increased from 3000 to 4500 m, while the cumulative gain of families was less than 10 and rarely changed with the elevation. This resulted in the reduction of the cumulative taxa richness shown in Fig. 7.7. Furthermore, the regional richness (S_r) composition of insect families and non-insect groups was compared for different elevation ranges to show the changes in the assemblage composition as the elevation increased (Fig. 7.9). In the entire basin, the most family-rich insect orders were Plecoptera, Diptera, Trichoptera, and Epemeroptera. These four groups did not vary consistently with elevations, instead, the regional taxa richness of most of the groups peaked at intermediate elevations. The regional taxa richness of Trichoptera peaked at elevation 4000 to 4500 m. The regional taxa richness of Plecoptera and Diptera were the highest at elevation 3500 to 4000 m. The regional taxa richness of Epemeroptera peaked at elevation 3000 to 3500 m. Odonata, Coleoptera, Hemiptera, Megaloptera, and Entomobryomerpha accounted for a minimal amount of family richness. However, they were scattered throughout of the basin at different elevations and were essential for maintaining a high regional biodiversity.

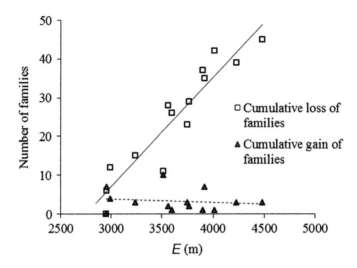

Figure 7.8 Relations between the cumulative loss of families, cumulative gain of families, and elevation.

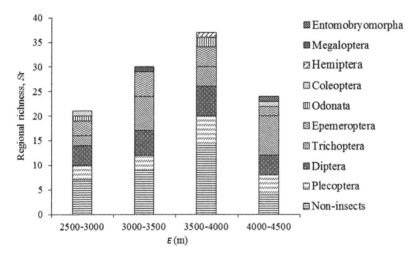

Figure 7.9 Regional richness compositions of different families at different elevation ranges.

7.1.5 Effect of environmental parameters on macroinvertebrate

Many alpine areas have high gradients, and, therefore, the climate, riparian vegetation, and water temperature vary drastically in these areas, all of which affect macroinvertebrate taxa composition (Jeník, 1997; Füreder et al., 2001). The authors performed a DCA with assistance of the software CANOCO for Windows Package 4.5 to explore the distributions of taxa at the sampling sites and the gradients of different environmental parameters. The densities of the 110 taxa identified at the 14 sampling sites were used as the input data in CANOCO 4.5. In a DCA analysis, the sampling sites with similar taxa composition are gathered in a group. In a DCA plot, Axis 1 indicates

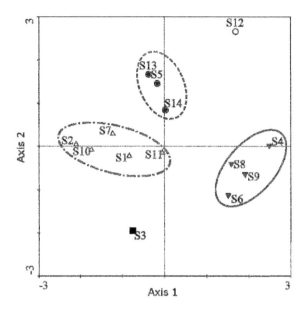

Figure 7.10 DCA plots of the sampling sites.

the ordination along the greatest gradient among the sampling sites; and Axis 2 indicates the ordination along the second greatest gradient among the sites. By this way, Fig. 7.10 shows the ordination diagram of the sampling sites, which were divided into five groups. In each group, the sampling sites shared similar composition of macroinvertebrate assemblages. The eigenvalues of Axes 1 and 2 of the DCA plots were 0.685 and 0.481, respectively, and explained 26.3% of the overall variance of the taxa composition among all of the sampling sites.

The variation of environmental parameters (Table 7.1) is believed to be the main reason for the variance in the taxa composition. For instance, the sampling sites characterized by variance of water temperature were mainly distributed along Axis 1. The sites with water temperatures below and above 10°C were separated along the left portion (S1, S2, S3, S5, S7, S10, S11) and right portion (S4, S6, S8, S9, S12, S14) of Axis 2, respectively. Milner and Petts (1994) defined glacier-fed rivers as the rivers with water temperatures below 10°C, and stated that the macroinvertebrate assemblages in glacier-fed rivers (water temperatures <10°C) were different from those in rainfall streams (water temperatures >10°C). Therefore, the varying effects of water temperature on macroinvertebrates may be caused by different sources of river water.

In addition, Fig. 7.10 also indicated that the bed structures and river patterns also affected sorting of the sampling sites along Axis 1. The sites with step-pool characteristics were gathered along the left portion of Axis 2, while the sites with braided river characteristics were along the right portion of Axis 2. The flow velocity also affected grouping of the sites along Axis 2. The sites with suitable flow velocities (0.3–0.8 m/s) were usually gathered around or below 0, while the sites with unsuitable flow velocities (<0.3 or >0.8 m/s) were gathered far higher than 0. Beauger et al. (2006) indicated that the riverbed tends to be filled and is not very productive with flow velocities below 0.3 m/s; therefore, taxa richness is low. In addition, the flow velocities above 1.2 m/s constrain most living materials and organisms.

Furthermore, riparian conditions, including riparian vegetation and bank structures, also affected the grouping of sites. For instance, the site S3 shared physical conditions similar to S6, S8, and S9. All were in a braided river with stable cobble beds and suitable flow velocities for macroinvertebrates. The only difference was that S3 had concrete levees while the other three had natural river banks. As a result, S3 was grouped far away from the group of S6, S8, and S9. The main reason was considered that the concrete levees of S3 destroyed the connectivity and integrity of macroinvertebrate habitat and consequently changed the assemblage composition.

In summary, Fig. 7.10 indicates that the water source, flow velocity, bed structure, river pattern, and riparian conditions all play important roles in structuring the macroinvertebrate assemblages in the Yarlung Tsangpo basin. Step-pool bed structures, natural stream banks, and suitable flow velocities are the most important parameters for sustaining the aquatic ecosystem.

In general, the taxa composition and density of the macroinvertebrates are very heterogeneous in the Yarlung Tsangpo basin, although the taxa richness varies slightly among different sites. The taxa composition of each site is only a small portion of the regional taxa composition of the entire basin. Therefore, the beta diversity of the Yarlung Tsangpo basin is higher than that of the rivers at lower-elevations with similar habitat conditions and alpha diversities.

7.2 AQUATIC ECOLOGY IN SANGJIANGYUAN

7.2.1 Ecological conditions in Sanjiangyuan

The area of Sangjiangyuan plays a key role in maintaining the unique biota and ecological balance of the river basins. However, the natural environment in the Sanjiangyuan region is very harsh, and the ecosystem is considered to be very fragile (Tang, 2003; Chen et al., 2007). Assessment of the ecological conditions of this region is a prerequisite for precautionary environmental management (Li et al., 2012).

In recent years, a number of studies have documented changing environmental conditions in the Sanjiangyuan region. These studies have involved climate (e.g. Li et al., 2006; Hu et al., 2007; Zhang et al., 2011), hydrology (e.g. Li et al., 2004; Zhang et al., 2004; Shi et al., 2007), terrestrial ecosystems (especially grassland ecosystems), grassland plant productivity (Sun et al., 2005; Guo et al., 2008), degradation and stress factors (Liu et al., 2008; Feng et al., 2009; Li et al., 2011; Yang et al., 2011), and sustainable utilization of grasslands (Sheng et al., 2007). Although spatial-temporal changes of habitat conditions for biotic organisms have been well documented (Qian et al., 2006; Wu et al., 2008; Liu et al., 2009), no systematic studies on aquatic biotic assemblages have been completed in the Sanjiangyuan region. Detailed analyses are required to select representative organisms as indicators of the aquatic ecological condition in this region.

Benthic macroinvertebrates are important components of river ecosystems. They play an important role in trophic dynamics by cycling nutrients and providing food for higher trophic level species such as fish and birds. The feeding function of benthic macroinvertebrates reflects the different types of materials that these organisms ingest and their predominant feeding behaviors (Smock, 1983). Generally, the benthic macroinvertebrates can be grouped into five functional feeding groups: shredders, scrapers, collector-filterers, collector-gatherers, and predators according to their feeding habits (Plafkin et al., 1989; Barbour et al., 1999). In addition, benthic macroinvertebrates typically live for quite a long time and have limited ability to migrate. Therefore, researchers have used benthic macroinvertebrates as indicators of long-term changes in the environment (Hart and Fuller, 1974; Smith et al., 1999; Pan et al., 2012). A better understanding of macroinvertebrate assemblages is very important in river ecological assessment and management.

Construction of hydroelectric power stations and wetland degradation have caused significant negative impacts upon aquatic ecology in the rivers. To date, no hydropower stations with significant water storage and flow regulation exist in the Sanjiangyuan region. However, the region has been subjected to significant wetland and grassland degradation owing to human activities such as overstocking, trenching and draining, illegal mining, and so on (Jiao et al., 2007; Wang et al., 2009).

A systematic investigation of the macroinvertebrates in the Sanjiangyuan region was carried out in 2009–2010. The purpose of the investigation was threefold: 1) to describe the overall characteristics of the macroinvertebrate assemblages in Sanjiangyuan; 2) to evaluate the impacts of human activities on the macroinvertebrate assemblages; and 3) to develop strategies for aquatic conservation and management in the Sanjiangyuan region. This chapter only discusses the ecological conditions in the source areas of the Yellow and Yangtze rivers. The ecological conditions in the source area of the Mekong river has not yet been investigated comprehensively owing to logistics difficulties.

7.2.2 Field investigations of macroinvertebrates

Fluvial morphology and river patterns in the source area of the Yellow River differ markedly from those in the source area of the Yangtze River. The source area of the Yellow River is at lower elevations and much more varied in geomorphology than the source area of the Yangtze River. The main channel of the Yellow River source area flows from elevation 4600 to 2500 m. Variation in the river bed gradient in the Yellow River source area is also more remarkable than that in the Yangtze River source area. The river patterns in the Yellow River source area include braided, anabranching, anastomosing, meandering and straight (Blue and Brierley, 2013; Li et al., 2013; Yu et al., 2013). In contrast, the Yangtze River source area is relatively flat (at elevation 4300–4800 m), divided by quasi-parallel low mountains or ridges, and the elevation gradually lowers from northwest to southeast. Wide and shallow valleys and braided channels are the dominant river patterns in the Yangtze River source area, although meandering and straight channel patterns also exist in laterally-confined reaches.

Field investigations of benthic macroinvertebrates in the Sanjiangyuan region were carried out in August 2009 and July 2010. The study locations in the Yellow River source region (33°46′01″–36°33′16″N, 97°45′22″–101°33′40″E) and the Yangtze River source region (33°43′56″–35°34′57″N, 92°07′05″–94°01′27″E) are shown in

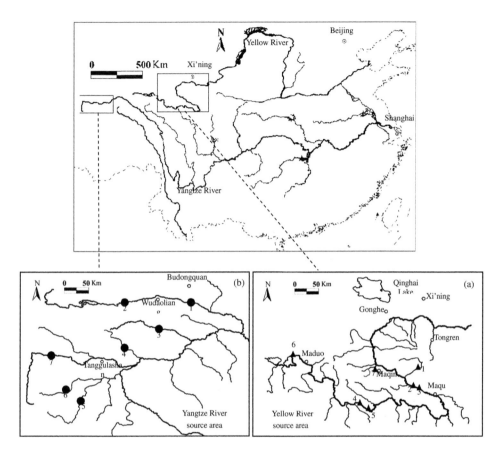

Figure 7.11 Study areas and sampling sites in the source areas of the Yellow River and the Yangtze River. (a) Sampling sites in the Yellow River source area: 1. Zequ; 2. a gravel bar along a meandering section of the Yellow River; 3. an oxbow lake adjacent to the Yellow River; 4. a riparian wetland by the Yellow River; 5. mainstream of the Yellow River (with low sediment concentration); 6. shore of the Erling Lake; and 7. the Jiangrang hydroelectric power station on a tributary of the Qiemuqu River. (b) Sampling sites in the Yangtze River source area: 1. downstream of the Chuma'er River; 2. upstream of the Chuma'er River; 3. Beilu River; 4. Ri'achiqu River; 5. Buqu River; 6. Ganaiqu River; and 7. Tuotuo River.

Fig. 7.11. Sampling and analyses of macroinvertebrates and environmental parameters were conducted using the same methods described in Section 7.1.1. The measured environmental parameters at the sampling sites were presented in Table 7.2.

The Jaccard similarity coefficient (S_J) was used to compare the macroinvertebrate assemblages in the two river source areas:

$$S_J = c/(a + b - c) \tag{7.4}$$

where a is the number of species in assemblage A, b is the number of species in assemblage B, and c is the number of species co-existing in both assemblages.

Table 7.2 Environmental parameters (mean ± SE) of sampling sites.

Environmental parameters	Yellow River source area	Yangtze River source area
Water depth (m)	0.4 ± 0.2	0.4 ± 0.2
Sechhi depth (m)	0.3 ± 0.1	0.15 ± 0.10
Water velocity (m/s)	0.26 ± 0.15	0.38 ± 0.05
Suspended sediments (mg/L)	59.3 ± 39.3	351.6 ± 84.5
Conductivity (μS/cm)	487 ± 12	2257 ± 838
Total nitrogen (mg/m^3)	2940 ± 540	1967 ± 347
Total phosphorus (mg/m^3)	8 ± 1	37 ± 17

In addition, taxa richness, S, the Shannon-Weaver index, H', and the improved Shannon-Weaver index, B were also analyzed using the same formula in section 7.1.1.

7.2.3 Taxa and biodiversity

A total of 68 taxa of macroinvertebrates belonging to 29 families and 59 genera were identified in the two river source areas (Table 7.3). Among these taxa, there were 8 annelids, 5 mollusks, 54 arthropods and 1 other animal. In the source area of the Yellow River, 50 taxa belonging to 25 families and 46 genera were identified. In the source area of the Yangtze River, 29 taxa belonging to 11 families and 24 genera were identified. Only 11 taxa of macroinvertebrates existed in both river source areas. The Jaccard coefficient between the source areas of the Yellow River and the Yangtze River was only 0.16, indicating that the similarity of macroinvertebrates between the two river source areas was low.

The Shannon-Weaver Index (H') of the macroinvertebrates in the source areas of the Yellow River and the Yangtze River was 2.06 and 2.05, respectively. However, the improved Shannon-Weaver index (B) of the macroinvertebrates was 11.86 and 8.41, respectively. Species diversity can also be assessed using K-dominant curves, which combine the two aspects of diversity, the species richness and evenness. Using this method, dominance patterns can be represented by plotting the accumulative abundance of each species (%) ranked in decreasing order of dominance. Figure 7.12 shows the K-dominant curve of macroinvertebrates in the river source areas. Biodiversity comparison indicated that the macroinvertebrate diversity in the source area of the Yellow River was higher than that in the source area of the Yangtze River.

7.2.4 Densities, biomass and functional structure of macroinvertebrates

Figure 7.13 shows the density and biomass of each taxonomic group of macroinvertebrates in the source areas of the Yellow River and the Yangtze River. The densities of the total macroinvertebrates were 329 ± 119 (mean ± SE) and 59 ± 32 individuals/m^2 in the source areas of the Yellow River and the Yangtze River, respectively. The biomass of the total macroinvertebrates was 0.3966 ± 0.1763 (mean ± SE) and 0.0307 ± 0.0217 g dry weight/m^2 in the source areas of the Yellow River and the Yangtze River, respectively. Arthropods were the predominant group in these two river source areas. In the

Table 7.3 Taxonomic composition of macroinvertebrates in the source areas of the Yellow River and the Yangtze River.

Phylum	Class	Family	Species (genus) number	
			The source region of the Yellow River	The source region of the Yangtze River
Nematoda			ud	0
Annelida	Oligochaeta	Naididae	I	0
		Tubificidae	4	4
Mollusca	Gastropoda	Lymnaeidae	3	0
		Planorbidae	2	0
Arthropoda	Crustacea	Gammaridae	ud	ud
	Arachnoida		ud	0
	Insecta	Caenidae	(I)	0
		Baetidae	(I)	(I)
		Heptageniidae	(I)	(I)
		Ephemerellidae	(I)	0
		Leptophlebiidae	0	(I)
		Hydropsychidae	(I)	0
		Leptoceridae	(I)	0
		Brachycentridae	(I)	0
		Nemouridae	0	ud
		Taeniopterygidae	0	ud
		Dytiscidae	ud	0
		Elmidae	ud	ud
		Chrysomelidae	ud	0
		Naucoridae	ud	0
		Corixidae	ud	0
		Pyralidae	ud	0
		Sialidae	0	ud
		Tipulidae	(I)	(I)
		Simuliidae	(I)	0
		Culicidae	ud	0
		Ephydridae	ud	0
		Chironomidae	(20)	(16)
		Total taxa number	50	29

Note: ud, taxon unidentified to genus or species; genus number in parentheses.

source area of the Yellow River, arthropods made up 85.4% of the total density and 93.5% of the total biomass. In the source area of the Yangtze River, arthropods comprised 86.2% of the total density and 95.8% of the total biomass.

Figure 7.14 shows the density and biomass of each functional feeding group of the macroinvertebrates in the source areas of the Yellow River and the Yangtze River. Among all the functional groups, shredders were the predominant group in the source area of the Yellow River, making up 65.7% of the total density and 87.1% of the total biomass. In the source area of the Yangtze River, collector-gatherers and shredders were the dominant groups. Collector-gatherers made up 62.0% of the total density and 50.6% of the total biomass, and shredders made up 28.2% of the total density and 40.7% of the total biomass.

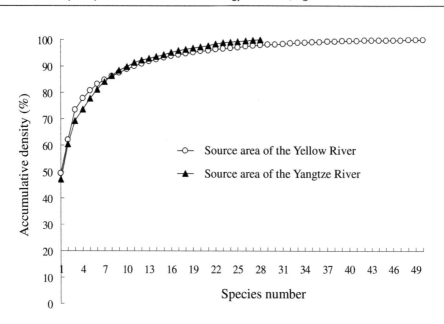

Figure 7.12   K-dominant curves of macroinvertebrates in the source areas of the Yellow River and the Yangtze River.

7.2.5   Effect of human disturbances

To assess the impact of hydropower station construction on aquatic ecology, macroin-vertebrate assemblages in the upstream and downstream reaches of the Jiangrang run-of-river hydroelectric power station in Maqin County were compared. Little vari-ance of taxa number, density and biomass of macroinvertebrates was identified for the upstream and downstream reaches of the hydroelectric power station (Table 7.4).

Macroinvertebrate assemblages in riparian wetland sites were more diverse than those in a tributary without macrophytes (Table 7.4), indicating that there were much more abundant benthic animal resources in the riparian wetlands than in the tributary. The total density and biomass of macroinvertebrates in the riparian wetlands were 21 and 27 times higher than those in the tributary without macrophytes. More details about influence of macrophytes on river ecology will be discussed in the following text.

In terms of the macroinvertebrate assemblage structure, aquatic insects were dom-inant in the source areas of the Yellow and Yangtze Rivers. Similar features have been found in the mid-lower reaches of the trunk streams (Xie et al., 1999; Pan et al., 2011). With regard to the species composition, some potamophilic and psychrophilic species were found in these two river source areas. The potamophilic taxa included Ephemeroptera, Simuliidae, *Rheotanytarsus, Stictochironomus, Xenochironomus,* etc. The psychrophilic species included *Stylaria lacustris, Limnodrilus grandisetosus, Limnodrilus profundicola, etc.* The existence of psychrophilic species is ascribed to the high elevations and low water temperature. The source area of the Yellow River was found to have higher macroinvertebrate diversity than the source of the Yangtze River, , which was believed to be ascribed to the lower elevations and better wetland development in the source area of the Yellow River. In general, environmental condi-tions such as climate, oxygen concentration, and atmospheric pressure are sufficient

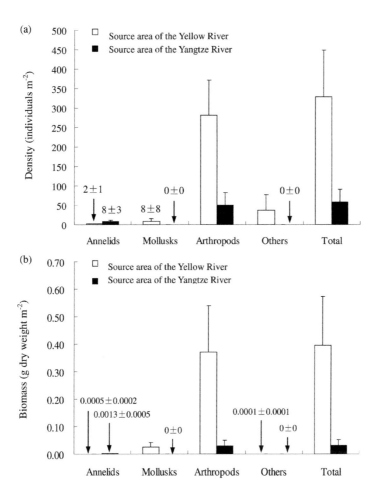

Figure 7.13 Mean (±SE) density (a) and biomass (b) of each taxonomic group of macroinvertebrates in the source areas of the Yellow River and the Yangtze River.

at lower elevations. In addition, better wetland development supports more benthic taxa, as these ecological features play an important role in the maintenance of biota diversity.

In areas with limited human disturbance, the macroinvertebrate assemblages are influenced primarily by habitat conditions such as water velocity (Brooks et al., 2005), substrate size (Jowett and Richardson, 1990; Quinn and Hickey, 1990), bank morphology, and other geomorphological factors (Armitage et al., 2001; Chessman et al., 2006). The variance of the macroinvertebrate assemblages in the source areas of the Yellow and Yangtze Rivers reflects different habitat conditions. Among all measured parameters, sediment concentrations in these two river source areas had the largest difference. The sediment concentration in the source area of the Yangtze River was six times of that in the source of the Yellow River (Table 7.2). To a large extent, the less abundant animal resources in the source region of the Yangtze River is ascribed to the high sediment concentration. High concentration of sediment blocks the light

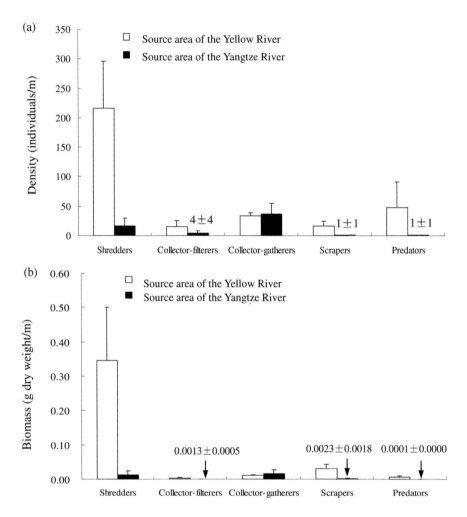

Figure 7.14 Mean (+SE) density (a) and biomass (b) of each functional feeding group of the macroinvertebrates in the source areas of the Yellow River and the Yangtze River.

Table 7.4 Comparisons of taxa number, densities and biomass of macroinvertebrates at different locations.

Study station	Taxa number	Density (individuals m^{-2})	Biomass (g dry weight m^{-2})
Upstream of the Jiangrang hydroelectric power station	10	59 ± 30	0.0100 ± 0.0053
Downstream of the Jiangrang hydroelectric power station	8	43 ± 31	0.0075 ± 0.0050
A tributary without macrophytes	7	11 ± 11	0.0051 ± 0.0051
Riparian wetlands	14	232 ± 92	0.1374 ± 0.0651

to support algal growth, and sand pellets may spoil cell walls of phytoplankton (Allan and Castillo, 2007; Pan et al., 2009), resulting in food shortage for benthic animals. High concentration of sediment also inhibits macrophyte growth by blocking the light and impacting the benthic dwellers as macrophytes can support a large number of epiphytic animals (Brönmark, 1989; Newman, 1991; Jeppesen et al., 1998; Wang et al., 2006). Macrophytes create significant horizontal and vertical heterogeneities that provide a physical condition for distinct niches (Rosine, 1955; Rooke, 1984) and serve as a refuge for epiphytic animals against their predators (Pan et al., 2011). Moreover, macrophytes can serve as sites for oviposition (Rooke, 1984; Scheffer, 2004) and provide chances for snails to crawl on the air-water interface (particularly for pulmonates) (Brönmark, 1989; Wang et al., 2006).

Flow regulation impacts in the two river source areas are limited to a few run-of-river hydroelectric power stations, one of which is shown to have insignificant impacts on benthic dwellers (Table 7.4). However, it has been noted in several studies that aquatic ecology upstream and downstream of non-river-of-river power stations may vary greatly due to the altered hydrological regime (Friedl and Wüest, 2002; Godlewska et al., 2003; Jorcin and Nagueira, 2005; Allan and Castillo, 2007). The limited impacts caused by the Jiangrang power station demonstrate that run-of-river hydroelectric power stations have less adverse impacts on the natural flow regime compared to non-river-of-river power stations. For instance, the Jiangrang run-of-river hydroelectric power station raises the headwater by less than 1 m, and its storage capacity is very limited. Thus, the run-of-river hydroelectric power stations have very limited impacts on the flow regime.

In addition to the construction of run-of-river hydroelectric power stations, the aquatic ecosystems in the river source areas are also suffering from wetland degradation caused by human activities. As noted in many other studies (e.g. Hillman and Quinn, 2002; Tarr et al., 2005; Sharma and Rawat, 2009), wetlands support much greater abundance of benthic individuals than habitats without vegetation (Table 7.4).

Spatial comparisons of the benthic assemblages and environments in these two river source areas have indicated that macroinvertebrates were mainly influenced by sediment concentration and aquatic vegetation. The taxa number, density and biomass of macroinvertebrates in the source area of the Yellow River were 1.7, 5.6 and 12.9 times of those in the source area of the Yangtze River. The less abundant benthic animal resources in the source of the Yangtze River were ascribed to higher elevations, higher sediment concentration and more severe wetland degradation. In recent years, grassland degradation and erosion has led to serious soil loss, increasing sediment concentration in this region (Luo and Tang, 2003). In the past 40 years, wetland areas in the source regions of the Yellow River and the Yangtze River have decreased by 13.6% and 28.9%, respectively (Wang et al., 2009). Degradation of wetlands is due to not only human activities but also reduced rainfall as a result of climate change. Conservative grazing and vegetation management can support efforts to minimize grassland and wetland degradation and desertification, and reduce soil erosion rate and river sediment discharge.

In summary, run-of-river hydroelectric power stations have limited impacts on benthic animals, while wetland degradation has significant impacts on macroinvertebrates. Vegetation management and re-instigation of wetlands in an effort to reduce soil erosion and maintain water supply are key measures to protect the ecological conditions in these source regions and improve the ecological conditions across the entire river basins.

7.3 AQUATIC ECOLOGY OF ABANDONED CHANNELS

Fluvial processes may result in channel abandonment. Abandoned channels are a very unique landscape and play a special role in the river ecosystem. There are four types of abandoned channels: old river courses, oxbow lakes, ox-tail lakes and riparian wetlands. These four types of abandoned channels result from avulsions, meander cutoffs, ice-jam floods and stem-channel shifts, respectively.

Avulsion is an inevitable result of river aggradation and is therefore closely related to the sediment load that the stream carries (Allen, 1965; Field, 2001). In heavy sediment-laden rivers, such as the Yellow River, avulsion is the dominant mechanism of channel shifting on alluvial fans and river deltas, resulting in devastating calamities and the creation of numerous old channels. River courses resulting from avulsion have no connection with the present river channel and have little influence on the river ecology.

Oxbow lakes result from natural cutoffs of river meanders. In meandering rivers, a continuing increase in the amplitude and tightness of bends may result in the river reaching a threshold sinuosity at which the river can no longer maintain its shape and a cutoff develops. Oxbow lakes may also result from artificial cutoffs. In general, artificial cutoffs cause intensive erosion in the new channel in the first several years of formation. The new channel is not stable during the intensive fluvial process.

Ox-tail lakes are generated from the fluvial process of anastomosing rivers (Wang et al., 2010). In northeastern China, some rivers flow from south to north. When the northern portion of the river freezes, the water from the south continues to flow to the north. The frozen reach keeps the water from continuing to flow in the original channel, and causes the water to erode the river bank and create new channels on the original floodplain. These new channels are called anastomosing rivers. Anastomosing rivers are not very stable. If one of the newly-created channels is scoured deeper than the others, all available water may flow into this channel and abandon the others. The abandoned channels remain connected with the main channel and form a channel-shaped lake. These lakes are called ox-tail lakes.

Many riparian wetlands result from the shifting of channels. The shifting of channels causes sediment to deposit in wide river sections and form sand bars. Under some circumstances, one channel of the braided river develops into the main channel, and the other channels and the bars turn into a wetland.

With the exception of old river courses, which belong to terrestrial ecosystems, the other types of abandoned channels, such as oxbow lakes, ox-tail lakes and riparian wetlands, are important components of freshwater ecosystems. These abandoned channels provide habitats for animals, and are important in maintaining the unique and complicated biota of river systems.

Previous studies on macroinvertebrates in abandoned channels were mainly focused on oxbow lakes instead of descriptions of assemblage structures (Richardot-Goulet et al., 1987; Brock and van der Velde, 1996; Pan et al., 2008). The lack of research on potential factors that influence benthic assemblages have impeded human's ability to conserve and manage abandoned channels. Some studies have revealed that the hydrological connectivity between river-associated water bodies and river mainstreams is a potential driving force for structuring benthic assemblages (Van den Brink and Van der Velde, 1991; Amoros and Bornette, 2002). The intermediate disturbance hypothesis was first reported by Connell (1978) as a mechanism that maintains high diversity in tropical forests and coral reefs. Recent literature has suggested that this mechanism may be much broader in its scope than previously recognized. Studies have shown that the α-diversity of macroinvertebrates peaks at an intermediate level of hydrological connectivity (Obrdlik and Fuchs, 1991; Tockner et al., 1999; Ward et al., 1999; Amoros and Bornette, 2002).

Hydrological connectivity mainly affects benthic assemblages by altering the substrate type, which plays an important role in the distribution and abundance of macroinvertebrates. The substrate characteristics which affect the resident macroinvertebrate assemblages include particle size, stability and heterogeneity. In general, the diversity of benthic invertebrates increases with the median particle size (Minshall and Minshall, 1977; Wise and Molles, 1979), while some evidence suggests that the diversity declines when particles are larger than the size of cobbles (Minshall, 1984). The diversity and density of macroinvertebrates are the lowest in unstable substrates (e.g. sand), which does not provide enough food or a sufficiently stable habitat for macroinvertebrates (Verdonschot, 2001; Jowett, 2003). Macroinvertebrates are the most abundant in stable substrates (Bunn and Davies, 1990; Hax and Golladay, 1998). Higher substrate heterogeneity supports more diverse benthic species (Erman and Erman, 1984; Gayraud and Philippe, 2001). Macrophytes are also considered to be a type of substrate. This is ascribed to the fact that macrophytes not only directly or indirectly provide food for benthic animals, but also increase the spatial heterogeneity (Brönmark, 1989; Jeppesen et al., 1998). Therefore, the role of macrophytes in habitat diversity should be considered (Wang et al., 2008).

7.3.1   Typical types of abandoned channels

Systematic investigations of macroinvertebrates were conducted by the authors of this book in three types of abandoned channels during 2006–2009. The water bodies studied that were investigated are situated in the source area of the Yellow River, Songhua River basin and East River basin. The source area of the Yellow River has a total drainage area of 131,420 km². The total area of wetlands in the Yellow River source is 38,000 km², including the river, riparian waters, lakes and swamps (Wang et al., 2010). The Songhua River is 2,309 km long, starting from the Neng River and ending at the Heilong River. The drainage area of the Songhua River basin is 5.568×10^6 km² (Liu et al., 2008), and includes many anastomosing rivers and ox-tail lakes. The East River is 562 km long and has a drainage area of 35,340 km². The river is one of three major rivers of the Pearl River system—the largest system in Southern China. The East River carries a light sediment load, averaging 2.96×10^6 t/yr (Wang et al., 2008).

The locations of the study areas and sampling sites are shown in Fig. 7.15. Hydrological characteristics and environmental conditions of the sampling sites are given in Table 7.5. Field investigations were conducted in the Yellow River source area in August 2009, in the Songhua River system and its tributary, the Mudan River, in

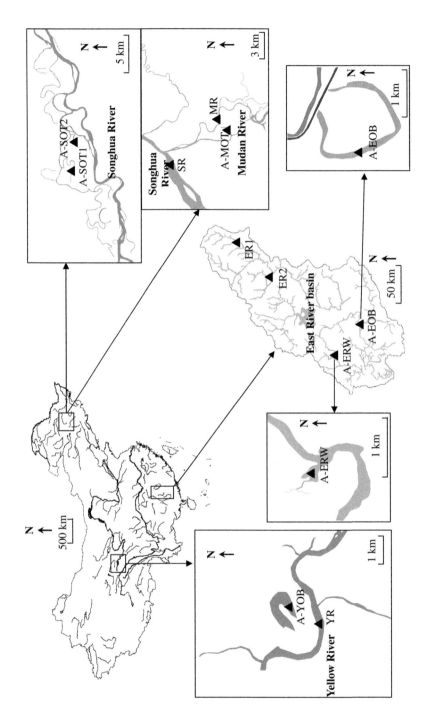

Figure 7.15 The locations of the study areas and sampling sites (The site abbreviations are defined in Table 7.5. ▲ represents the sampling site).

Table 7.5 Hydrological characteristics and environmental conditions of the sampling sites.

Site		Code	Location	Connection frequency (yr⁻¹)	Substrate	Velocity (m s⁻¹)	Water depth (m)	Method for frequency determination	Effect of human activities
Yellow River	Oxbow lake of the Yellow River at Kesheng	A-YOB	N 34°12′4″ E 101°33′40″	0.3	Clay and silt, dense emerged and submerged plants (cover: 1/3)	0.0–0.3	0.3–1.0	Interviews with local residents	Little effect
	Mainstream of the Yellow River at Kesheng	YR	N 34°12′3″ E 101°33′39″		Silt, sand and cobbles	0.1–1.0	0.1–1.0		
Songhua River	The first ox-tail lake of the Songhua River in Harbin	A-SOT1	N 45°47′1″ E 126°23′25″	1.0	Silt and fine sand, aquatic plant (cover: 1/3)	0.0–0.2	0.1–1.5	Hydrologic records	Separated by levee and regular open lock
	The second ox-tail lake of the Songhua River at Wanbao	A-SOT2	N 45°47′58″ E 126°32′18″	0.1	Silt and fine sand	0.0–0.2	0.1–1.5	Interviews with local residents	
	Mainstream of the Songhua River at Yilan	SR	N 45°47′5″ E 126°36′37″		Sand and gravel	0.1–0.8	0.1–3.0		
Mudan River	Ox tail lake of the Mudanjiang River at Xiaowujia	A-MOT	N 45°49′5″ E 126°44′4″	1.0	Clay and silt	0.0–0.2	0.1–0.6	Hydrologic records	Naturally connected the river during flood every year
	Mainstream of the Mudanjiang River at Xiaowujia	MR	N 45°49′4″ E 126°43′59″		Sand and gravel	0.3–0.8	0.1–1.5		
East River	Oxbow lake of the East River	A-EOB	N 23°3′23″ E 114°25′34″	0.0	Fine sand	0.0	0.0–3.0	Hydrologic records	Separated by levee and highway, which are never flooded
	Zengjiang Bay (riparian wetland of the East River)	A-ERW	N 23°27′1″ E 113°54′8″	Always connected	Clay, silt and sand, dense submerged plants (cover: 1/2)	0.0–0.5	0.0–3.0	Hydrologic records	
	First site in the mainstream of the East River at Shangpingshui	ER1	N 24°34′15″ E 115°29′46″		Cobbles and boulders, aquatic plants (cover: 1/4)	0.2–1.0	0.1–0.5		
	Second site in the mainstream of the East River at Yidu	ER2	N 24°17′36″ E 115°7′49″		Cobbles	0.2–1.5	0.2–1.0		

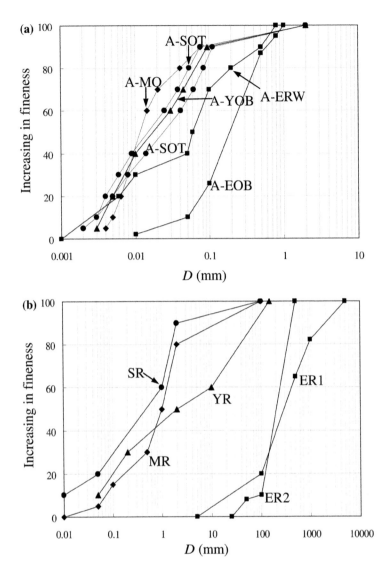

Figure 7.16 Size distributions of bed sediment at the sampling sites (Site abbreviations are defined in Table 7.5).

September 2009, and in the East River system in July 2006. Bed sediments samples were taken in the hyporheic zones of these rivers (depth = 0.15 m). The particle size distribution was analyzed using laser diffraction particle sizing (Malvern Mastersizer 2000). Figure 7.16 shows the size distributions of the bed sediment from the sampling sites. The biodiversity of macroinvertebrates was assessed based on the Shannon-Wiener index (H′), and the improved Shannon-Wiener index (B). The software STATISTICA 6.0 was used for analyses of simple regression. To reduce heterogeneity of variances, data were log 10-transformed.

7.3.2 Taxa composition and biodiversity

A total of 93 taxa of macroinvertebrates belonging to 51 families and 88 genera were identified form the samples collected (Table 7.6). Among them were 13 annelids, 17 mollusks, 61 arthropods and 2 miscellaneous animals. Only 10–30% of the taxa existed in both the abandoned channels and their adjoining rivers, which indicated that the abandoned channels had their own unique taxa composition. Oligochaetes were frequently found in the ox-tail lakes where the bottom was largely silt. Epiphytic taxa (e.g. Bithyniidae or Lymnaeidae) were found in the oxbow lakes of the Yellow River and in the riparian wetlands of the East River where dense macrophytes existed. Table 7.6 shows the biodiversity of the macroinvertebrates in the abandoned channels and their adjoining rivers. The riparian wetlands of the East River have the largest macroinvertebrate diversity. These riparian wetlands are freely connected with the mainstream of the East River.

The species diversity can also be assessed using K-dominant curves, which combines the two aspects of diversity—species richness and evenness. Using this method, dominance patterns can be represented by plotting the accumulative abundance of each species (%) ranked in a decreasing order of dominance. Figure 7.17 shows the K-dominant curves of the macroinvertebrates in the abandoned channels and their adjoining rivers. Figure 7.17 reports similar results as shown in Table 7.6.

Figure 7.18 shows the densities and biomass of each taxonomic group of the macroinvertebrates in the abandoned channels and their adjoining rivers. The total density peaked in the ox-tail lake of the Mudan River, where Chironomidae and Tubificidae were the most abundant. The total biomass peaked in the riparian wetlands of the East River, where mollusks were the predominant group. The standing crop in the abandoned channels that were covered by a layer of silt (e.g. the oxbow lake of the Yellow River and the ox-tail lakes of the Songhua River) was higher than that in the mainstreams, where the streambed sediment consist mainly of sand. The standing crop in the abandoned channels covered with fine sand (e.g. the oxbow lake of East River) was lower than that in the channel with gravel streambed.

Figure 7.19 shows the densities and biomass of each functional feeding group of the macroinvertebrates in the abandoned channels and their adjoining rivers. In the ox-tail lake of the Mudan River, which had the highest density, macroinvertebrates were predominately collector-gatherers. In the riparian wetlands of the East River, which had the highest biomass, macroinvertebrates were predominately scrapers.

Figure 7.20 shows the relations between the median particle size and the taxa number, density and biomass of the macroinvertebrates in the abandoned channels and their adjoining rivers. Because the median particle size in the abandoned channels differed from the median particle size in their adjoining rivers, regression analyses of the abandoned channels and the adjoining rivers were performed separately. Figure 7.20(a) shows that the taxa number of the macroinvertebrates increases initially and then decreases as the median particle size increases in the abandoned channels and their adjoining rivers. Figure 7.20(b) shows that as the median particle size increases, densities of the macroinvertebrates in the abandoned channels decrease, while densities in the river mainstreams increases initially and then decreases. Figure 7.20(c) reveals that as the median particle size increases, the biomass of the macroinvertebrates in the abandoned channels increases initially and then decreases, while the biomass in the river mainstreams always increases.

Table 7.6 Taxonomic composition and biodiversity of macroinvertebrates in abandoned channels and their adjoining rivers.

Phylum	Class	Family	Species (genus) number										
			A-YOB	YR	A-SOT1	A-SOT2	SR	A-MOT	MR	A-EOB	A-ERW	ER1	ER2
Platyhelminthes	Turbellaria		ud			ud					ud		
Nematoda													
Annelida	Oligochaeta	Naididae	1		2						1		
		Tubificidae			3	2	2	2			2		
	Hirudinea	Glossiphonidae			1		1	2	1				
		Herpodellidae			1			1	1		1	1	
		Hirudinidae			1								
	Mollusca Gastropoda	Viviparidae			1						2		
		Bithyniidae	2		2			1			1		1
		Lymnaeidae				1					1		
		Ampullariidae									1		
		Planorbidae	2		2			2			1		
		Pleuroseridae					1		1		1	1	1
	Bivalvia	Corbiculidae				1					1	1	1
		Mytilidae										1	
	Crustacea	Palaemonidae	ud	ud						1	1		1
	Arachnoida		ud	ud			ud						
	Insecta	Caenidae	1	1				ud			1	1	1
		Baetidae		1	1						1	1	
		Heptageniidae		1	1		1				1		
		Ephemeridae		1							1		
		Ephemerellidae							1				
		Leptophlebiidae											2
		Siphlonuridae											1
		Hydropsychidae		1								2	1
		Brachycentridae		1									1

Taxon										
Polycentropodidae									—	
Libellulidae								—		—
Gomphidae										ud
Macromiidae								ud		
Coenagrionidae					ud			—		
Platycnemididae								—		
Dytiscidae	ud		ud					—		
Elmidae	ud		ud						ud	
Hydrophilidae								—		
Psephenidae								—		
Hydrophilidae										
Chrysomelidae			2				—			
Naucoridae		ud			ud			ud		ud
Corixidae	ud				ud			ud		
Pyralidae	ud									—
Corydalidae					—					
Ceratopogonidae								—	—	
Tabanidae		ud						ud		
Stratiomyiidae					ud					
Tipulidae		—								
Simuliidae		—								
Culicidae			—		—				—	
Chironomidae	7	7	3	7	3	4	2	3	4	2	—
Total taxa number	20	17	20	15	9	16	6	3	30	13	14
H'	1.74	2.08	1.45	0.71	0.93	1.42	1.68	0.89	2.57	1.97	1.66
B	11.86	9.86	7.58	4.26	4.08	9.93	3.68	2.76	15.02	8.92	7.41

Note: Site abbreviations are defined in Table 7.5, and ud = taxon unidentified to genus or species.

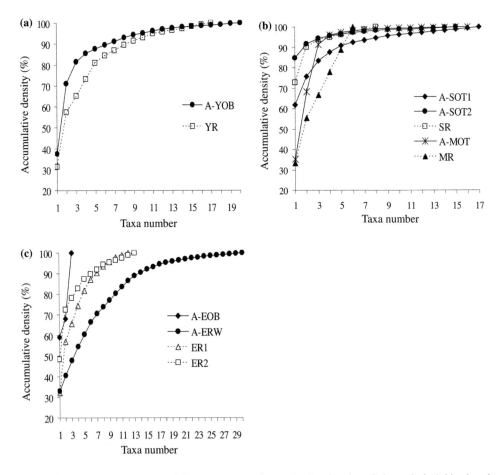

Figure 7.17 K-dominant curves of the macroinvertebrates in the abandoned channels (solid line) and their adjoining rivers (dashed line). The site abbreviations are defined in Table 7.5.

It is apparent in Fig. 7.20 that the taxa number of macroinvertebrates is the highest in the riparian wetlands of the East River. This is because the moderate hydrological connectivity has led to high heterogeneity, i.e., substrate diversity (Ward et al., 1999; Amoros and Bornette, 2002). It is also because that aquatic plants exist in the regions with slow flows, and these plants can increase spatial niches and living space for benthic animals (Brönmark, 1989; Jeppesen et al., 1998). However, silt and sand with low heterogeneity support fewer taxa. Therefore, taxa numbers show a unimodal change along the gradient of median particle size (Fig. 7.20a). The total macroinvertebrate density is the highest in the ox-tail lake of the Mudan River, where the benthic animals are dominantly collector-gatherers, which prefer soft sediments rich in organic matters (Strayer, 1985; Brinkhurst and Gelder, 1991). The downward trend of the density of benthic animals in the abandoned channel (Fig. 7.20b) is likely attributed to the fact that organic-rich silt sediment attract a large number of oligochaetes and chironomid larvae. The biomass of total macroinvertebrates is the greatest in the riparian wetlands

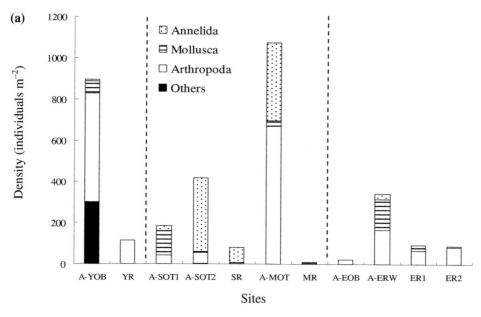

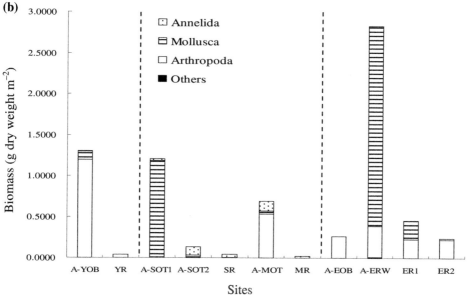

Figure 7.18 Densities and biomass of each taxonomic group of macroinvertebrates in the abandoned channels and their adjoining rivers. The site abbreviations are defined in Table 7.5.

of the East River, where benthic animals are dominantly scrapers (mainly Bellamya spp.) and collector-filterers (mainly Corbicula fluminea). The upward trend of the biomass of benthic animals in the abandoned channel (Fig. 7.20c) is attributed to the existence of large scrapers that prefer pebbles and cobbles.

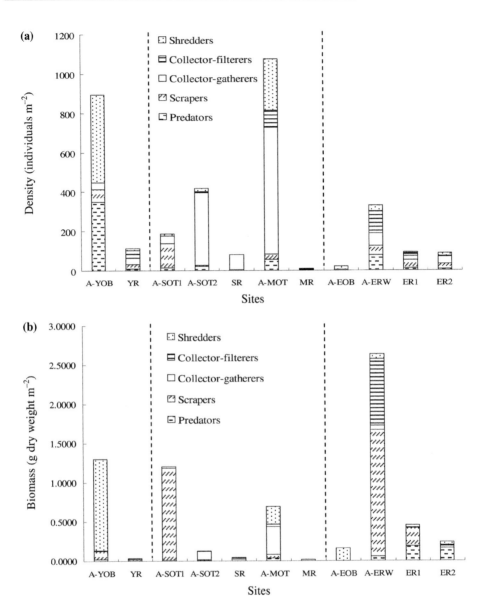

Figure 7.19 Densities and biomass of each functional feeding group of macroinvertebrates in the abandoned channels and their adjoining rivers. The site abbreviations are defined in Table 7.5.

In assemblage composition, most taxa in abandoned channels are similar to both river-isolated lakes and river mainstreams. Collector-gatherers (mainly Chironomidae and Tubificidae) which prefer soft sediment are the most abundant in some abandoned channels. Similar features have also been found in river-isolated lakes.

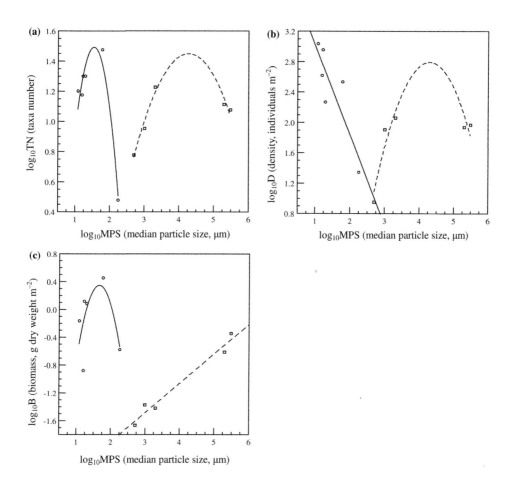

Figure 7.20 Relations between median particle size and taxa number (a), density (b), and biomass (c) of macroinvertebrates in the abandoned channels (solid line) and their adjoining rivers (dashed line). The site abbreviations are defined in Table 7.5.

Moreover, potamophilic taxa (e.g. Leptophlebia spp., Cinygmula spp., Brachycentrus spp., Neureclipsis spp., Stratiomyiidae, Simulium spp., Stictochironomus spp.) mainly exist in the abandoned channels that still connect with rivers. This is due to the exchange of nutrients and biota between the abandoned channels and the river mainstreams (Heiler et al., 1995; Ward et al., 1999).

The special hydrological conditions of abandoned channels provide important environment for the existence of a few unique benthic taxa, which are important resources for the river system. In the macrophytic regions of abandoned channels, unique taxa (i.e. some epiphytic and predatory taxa) are always supported. This is attributed to the fact that rich food sources of these taxa are either directly or indirectly provided by macrophytes (Gayraud and Philippe, 2001; Allan and Castillo, 2007).

The ranges of the sediment particle size in abandoned channels are significantly different from that in the main channels (see Fig. 7.16), which are ascribed to the variance in the hydrological connectivity. The connection frequency and the distance from the river to the abandoned channel are directly related to the extent of the hydrological connectivity. There are two reasons for the smaller median particle size in abandoned channels: 1) the increased distance from the river to the abandoned channel, which results in lower fluvial flow and leads to the deposition of suspended sediments; and 2) more frequent connections between the abandoned channel and the mainstream are conducive to the accumulation of finer sediments (Schwarz et al., 1996). Larger median particle size is also found in abandoned channels, ascribing to a rapid water current, which results from the direct interconnection between the abandoned channel and the mainstream (Hicks and Gomez, 2003; Allan and Castillo, 2007).

Substrate provides the primary living condition for benthic animals. Different types of substrate support different faunas. The substrate mainly affects benthic macroinvertebrates through its compactness, stability and heterogeneity (Angradi, 1999; Kaller and Hartman, 2004). Sand is generally considered to be the poorest substrate, especially for macroinvertebrates, due to its instability and detritus shortage (Allan and Castillo, 2007). In contrast, organic-rich silt sediment can sustain a large number of oligochaetes and chironomid larvae. Gravel beds support many insects because more stable microhabitats are formed in this type of bed. Therefore, standing crops in the abandoned channels which are covered with a layer of silt are higher than that in the river mainstreams which have bed sediment consisting mainly of sand. However, standing crops in the abandoned channels covered with fine sand are lower than that in the rivers with gravel streambed.

7.3.3 Management of abandoned channels

Abandoned channels provide unique habitats for animals which need to be protected. Previous studies have shown that it is necessary to conserve abandoned channels by restoring the natural hydrological connectivity (Paillex et al., 2007; Gallardo et al., 2008). In abandoned channels, the sediment deposition is closely related to the hydrological connectivity and determines the characteristics of the substrate, which play an important role in structuring the benthic assemblages. The riparian wetlands, which are freely connected with the mainstream, have high heterogeneity and the highest macroinvertebrate biodiversity and biomass. Previous research (Obrdlik and Fuchs, 1991; Ward, 1998; Tockner et al., 1999; Ward et al., 1999; Amoros and Bornette, 2002) has shown that the α-diversity of macroinvertebrates also peaks in the water bodies with moderate connectivity with rivers. On the contrary, rapid flow in the rivers scours the bed substrate, which is not conducive to the survival of benthic animals.

Hydrological connectivity is very important in structuring assemblages of macroinvertebrates in abandoned channels. Reconnecting side-arm branches, increasing connection frequency and flow duration time can increase the turnover rate between associated channels and the mainstreams, and, thus, increase the diversity and standing crops of macroinvertebrates. In order to maintain at least half of the maximum number of macroinvertebrate resources in the regions less affected by human activities, abandoned channels need to be reconnected with the mainstreams during flooding at least once every three years (e.g. the oxbow lake of the source area of the Yellow

River). In the regions more affected by human activities, the abandoned channels need to be connected with the mainstreams at least once every year during flooding (e.g. the ox-tail lakes of the Songhua River and those of the Mudan River). When the interconnection occurs, both ends of the abandoned channels should be connected with the mainstream as much as possible to allow sufficient exchange of nutrients and biota between the abandoned channels and the river mainstream.

REFERENCES

Allan, J.D. and Castillo, M.M. 2007. Stream Ecology: Structure and Function of Running Waters. AA Dordrecht: Springer.

Allen, J.R.L. 1965. A review of the origin and characteristics of recent alluvial sediments. Sedimentology 5: 89–101.

Amoros, C. and Bornette, G. 2002. Connectivity and biocomplexity in waterbodies of riverine floodplains. *Freshwater Biology* 47: 761–776.

Angradi, T.R. 1999. Fine sediment and macroinvertebrate assemblages in Appalachian streams: a field experiment with biomonitoring applications. *Journal of the North American Benthological Society* 18: 49–66.

APHA (American Public Health Association), 2002. Standard Methods for the Examination of Water and Wastewater. Washington DC: APHA.

Armitage, P.D., Lattmann, K., Kneebone, N. and Harris, I. 2001. Bank profile and structure as determinants of macroinvertebrate assemblages–seasonal changes and management. Regulated Rivers: Research and Management, 17: 543–556.

Banse, K. and Mosher, S. 1980. Adult body mass and annual production/biomass relationships of field population. Ecological Monographs 50: 355–379.

Barbour, M.T., Gerritsen, B.D. and Snyder, B.D. 1999. Benthic macroinvertebrate protocols. In Barbour M.T., Gerritsen, B.D., Snyder, B.D., Stribling, J.B. eds. Rapid Bioassessment Protocols for Use in Streams and Wadeable Rivers: Periphyton, Benthic Macroinvertebrates and Fish. EPA 841-B-99-002. U.S. Environmental Protection Agency; Office of Water, Washington, 7.1–7.20.

Beauger, A., Lair, N., Reyes-Marchant, P. and Peiry, J.L. 2006. The distribution of macroinvertebrate assemblages in a reach of the River Allier (France), in relation to riverbed characteristics. Hydrobiologia, 571(1): 63–76.

Blue, B. and Brierley, G.J. 2013. Geodiversity on the Upper Yellow River. Journal of Geographical Sciences, 23(5): 775–792.

Boyero, L. 2005. Multiscale variation in the functional composition of stream macroinvertebrate communities in low-order mountain streams. Limnetica, 24(3–4): 245–250.

Brinkhurst, R.O. and Gelder, S.R. 1991. Annelida: Oligochaeta and Branchiobdcllidi. In *Ecology and Classification of North American Freshwater Invertebrate*, Thorp, J.H., Covich, A.P. (eds). Academic Press: Inc., San Diego, CA; 400–433.

Brittain, J.E. and Milner, A.M. 2001. Ecology of glacier-fed rivers: current status and concepts. Freshw Biol, 46(12): 1571–1578.

Brock, T.C.M. and Van der Velde, G. 1996. Aquatic macroinvertebrate community structure of a *Nymphoides peltata*-dominated and macrophyte-free site in an oxbow lake. *Netherlands Journal of Aquatic Ecology* 30(2–3): 151–163.

Brönmark, C. 1989. Interactions between epiphytes, macrophytes and freshwater snail: a review. Journal of Molluscan Studies, 55(2): 299–311.

Brooks, A.J., Haeusler, T., Reinfelds, I. and Williams, S. 2005. Hydraulic microhabitats and the distribution of macroinvertebrate assemblages in riffles. Freshwater Biology, 50: 331–344.

Bunn, S.E. and Davies, P.M. 1990. Why is the stream fauna of southwestern Australia so impoverished? *Hydrobiologia* 194: 169–176.

Burgherr, P. and Ward, J.V. 2001. Longitudinal and seasonal distribution patterns of the benthic fauna of an alpine glacial stream (Val Roseg, Swiss Alps). Freshw Biol, 46(12): 1705–1721 doi:10.1046/j.1365-2427.2001.00853.x

Castella, C. and Amoros, C. 1988. Freshwater macroinvertebrates as functional describers of the dynamics of former river beds. *Verhandlungen Internationale Vereinigung für Theoretische und Angewandte Limnologie* 23: 1299–1305.

Chen, G.C., Chen, X.Q. and Gou, X.J. 2007. Ecological Environment in Sanjiangyuan Natural Reserve. Xining, China: Qinghai People's Press (in Chinese).

Chessman, B.C., Fryirs, K.A. and Brierley, G. 2006. Linking geomorphic character, behaviour and condition to fluvial biodiversity: implications for river rehabilitation. Aquatic Conservation: Marine and Freshwater Ecosystems, 16: 267–288.

Čiamporová-Zaťovčová Z., Hamerlík, L., Šporka, F. and Bitušík, P. 2010. Littoral benthic macroinvertebrates of alpine lakes (Tatra Mts) along an altitudinal gradient: a basis for climate change assessment. Hydrobiologia, 648(1): 19–34 doi:10.1007/s10750-010-0139-5

Duan, X.H., Wang, Z.Y. and Xu, M.Z. 2011. Effects of fluvial processes and human activities on stream macro-invertebrates. Int J Sediment Res, 26(4): 416–430 doi:10.1016/S1001-6279(12)60002-X

Duan, X.H., Wang, Z.Y., Xu, M.Z. and Zhang, K. 2009. Effect of streambed sediment on benthic ecology. Int J Sediment Res, 24(3): 325–338 doi:10.1016/S1001-6279(10)60007-8

Erman, D.C. and Erman, N.A. 1984. The response of stream invertebrates to substrate size and heterogeneity. *Hydrobiologia* 108: 75–82.

Feng, Y.Z., Yang, G.H., Wang, D.X. et al., 2009. Ecological stress in grassland ecosystems in source regions of Yangtze, Yellow and Lancang Rivers over last 40 years. Acta Ecologia Sinica, 29(1): 492–498. (in Chinese)

Field, J. 2001. Channel avulsion on alluvial fans in southern Arizona. *Geomorphology* 37: 91–104.

Fjellheim, A., Raddum, G.G., Vandvik, V. and Stunchlik E. 2009. Diversity and distribution patterns of benthic invertebrates along alpine gradients. A study of remote European freshwater lakes. Adv Limnol, 62: 167–190

Flecker, A.S. and Allan, J.D. 1984. The importance of predation, substrate and spatial rufugia in determining lotic insect distributions. *Oecologia* 64(3): 306–313.

Friedl, G. and Wüest, A. 2002. Disrupting biogeochemical cycles – consequences of damming. Aquatic Science, 64: 55–65.

Füreder, L., Schütz, C., Wallinger, M. and Burger, R. 2001. Physico-chemistry and aquatic insects of a glacier-fed and a spring-fed alpine stream. Freshw Biol, 46(12): 1673–1690 doi:10.1046/j.1365-2427.2001.00862.x

Gallardo, B., García, M., Cabezas, Á., González, E., González, M., Ciancarelli, C. and Comín, F.A. 2008. Macroinvertebrate patterns along environmental gradients and hydrological connectivity within a regulated river-floodplain. *Aquatic Science* 70: 248–258.

Gayraud, S. and Philippe, M. 2001. Does subsurface interstitial space influence general features and morphological traits of the benthic macroinvertebrate community in streams? Archiv für Hydrobiologie 151: 667–686.

Godlewska, M., Mazurkiewicz-Borón, G., Pociecha, A. et al., 2003. Effects of flood on the functioning of the Dobczyce reservoir ecosystem. Hydrobiologia, 504: 305–313.

Guo, L.Y., Wu, R., Wang, Q.C. and Ji, S.M. 2008. Influence of climate change on grassland productivity in Xinghai County in the source regions of Yangtze River. Chinese Journal of Grassland, 32(2): 5–10. (in Chinese)

Hart, C.W. and Fuller, S.L.H. 1974. Pollution Ecology of Freshwater Invertebrates. New York: Academic Press.

Hax, C.L. and Golladay, S.W. 1998. Flow disturbance of macroinvertebrates inhabiting sediments and woody debris in a prairie stream. *American Midland Naturalist* 139: 210–223.

Heiler, G., Hein, T. and Schiemer, F. 1995. Hydrological connectivity and flood pulses as the central aspects for the integrity of a river-floodplain system. Regulated Rivers: Research & Management 11: 351–361.

Henriques-Oliveira, A.L. and Nessimian, J.L. 2010. Aquatic macroinvertebrate diversity and composition in streams along an altitudinal gradient in Southeastern Brazil. Biota Neotropica, 10(3): 115–128 doi:10.1590/S1676-06032010000300012

Hicks, D.M. and Gomez, B. 2003. Sediment transport. In *Tools in Fluvial Geomorphology*, Kondolf GM, Piégay H (eds). Wiley: West Sussex; 425–461.

Hill, M.O. 1979. DECORANA – A Fortran Program for Detrended Correspondence Analysis and Reciprocal Averaging. Ecology and Systematics, Cornell University, Ithaca, New York.

Hillman, T.J. and Quinn, G.P. 2002. Temporal changes in macroinvertebrate assemblages following experimental flooding in permanent and temporary wetlands in an Australian floodplain forest. River research and Applications, 18(2): 137–154.

Hu, L.W., Yang, G.H., Feng, Y.Z. and Ren, G.X. 2007. Tendency research of climate warm-dry in source regions of the Yangtze River, the Yellow River and the Lantsang River. Journal of Northwest Sci-Tech University of Agriculture and Forest, 35(7): 141–146. (in Chinese)

Jacobsen, D. 2004. Contrasting patterns in local and zonal family richness of stream invertebrates along an Andean altitudinal gradient. Freshw Biol, 49(10): 1293–1305 doi:10.1111/j.1365–2427.2004.01274.x

Jeník, J. 1997. The diversity of mountain life. In: Messerli, B., Ives, J.D. eds. Mountains of the World. New York: Parthenon Publishing Group, 199–231

Jeppesen, E., Søndergaard, M.A., Søndergaard, M.O. and Christophersen, K. 1998. *The structuring role of submersed macrophytes in lakes*. Springer-Verlag: Berlin Heidelberg.

Jiao, J.C., Yang, W.Q., Zhong, X. et al., 2007. Factors of restoration in Ruoergai wetland and its conservation strategies. Journal of Sichuan Forestry Science and Technology, 28(1): 98–102. (in Chinese)

Jorcin, A. and Nagueira, M.G. 2005. Temporal and spatial patterns based on sediment and sediment-water interface characteristics along a cascade of reservoirs (Paranapanema River, South-east Brazil). Lakes and Reservoirs: Research and Management, 10: 1–12.

Jowett, I.G. and Richardson, J. 1990. Microhabitat preferences of benthic invertebrates in a New Zealand river and the development of in-stream flow-habitat models for Deleatidium spp. New Zealand Journal of Marine and Freshwater Research, 24: 19–30.

Jowett, I.G. 2003. Hydraulic constraints on habitat suitability for benthic invertebrates in gravel-bed rivers. *River Research and Applications* 19: 495–507.

Kaller, M.D. and Hartman, K.J. 2004. Evidence of a threshold level of fine sediment accumulation for altering benthic macroinvertebrate communities. *Hydrobiologia* 518: 95–104.

Krebs, C.J. 1978. Ecology: *The Experimental Analysis of Distribution and Abundance* (4th Edition). Harper and Row: New York.

Li, D.F., Tian, Y. and Liu, C.M. 2004. Impact of land-cover and climate changes on runoff of the source regions of the Yellow River. Journal of Geographical Sciences, 14(3): 330–338.

Li, L., Li, F.X., Guo, A.H. and Zhu, X.D. 2006. Study on the climate change trend and its catastrophe over "Sanjiangyuan" Region in recent 43 years. Journal of Natural Resources, 21(1): 79–85. (in Chinese)

Li, X.L., Brierley, G., Shi, D.J., Xie, Y.L. and Sun, H.Q. 2012. Ecological protection and restoration in Sanjiangyuan Natural Reserve, Qinghai Province, China. In: Higgitt, D. (Ed), Perspectives on Environmental Management and Technology in Asian River Basins. SpringerBriefs in Geography, Netherlands: Springer, 93–120.

Li, X.L., Gao, J., Brierley, G. et al., 2011. Rangeland degradation on the Qinghai-Tibet Plateau: Implications for rehabilitation. Land Degradation and Development, 22: 1–9.

Li, Z., Wang, Z., Pan, B., Du, J., Brierley, G., Yu, G. and Blue, B. 2013. Analysis of controls upon channel planform at the First Great Bend of the Upper Yellow River, Qinghai-Tibetan plateau. Journal of Geographical Sciences, 23(5): 833–848.

Liang, Y.L. and Wang, H.Z. 1999. Zoobenthos. In *Advanced Hydrobiology*, Liu, J.K. (ed). Science Press: Beijing; 241–259 (in Chinese).

Liu, J.Y., Xu, X.L. and Shao, Q.Q. 2008. The spatial and temporal characteristics of grassland degradation in the three-river headwaters region in Qinghai Province. Acta Geographica Sinica, 63(4): 364–376. (in Chinese)

Liu, R.P., Liu, H.J., Wan, D.J. and Yang, M. 2008. Characterization of the Songhua River sediments and evaluation of their adsorption behavior for nitrobenzene. Journal of Environmental Sciences 20: 796–802 (in Chinese with English abstract).

Liu, S.Y., Zhang, Y., Zhang, Y.S. et al., 2009. Estimation of glacier runoff and future trends in the Yangtze River source region, China. Journal of Glaciology, 55(190): 353–362.

Liu, Z.F., Tian, L.D., Yao, T.D., Gong, T.L., Yin, C.L. and Yu, W.S. 2007. Variations of 18O in Precipitation of the Yarlung Zangbo River Basin. Acta Geogr Sin, 62(5): 510–517 (in Chinese)

Lu, X.X., Zhang, S.R., Xu, J.C. and Merz, J. 2011. The changing sediment loads of the Hindu Kush-Himalayan rivers: an overview. In: Proceedings of the ICCE Workshop held at Hyderabad, India, September 2009

Luo, X.Y. and Tang, W.J. 2003. Eco-environmental problems at source area of the Yangtze River and countermeasures for their prevention and treatment. Journal of Yangtze River Scientific Research Institute, 20(1): 47–49. (in Chinese)

McGregor, G., Petts, G.E., Gurnell, A.M. and Milner, A.M. 1995. Sensitivity of alpine stream ecosystems to climatic change and human impacts. Aquat Conserv, 5(3): 233–247 doi:10.1002/aqc.3270050306

Milner, A.M., Brittain, J.E., Brown, L.E. and Hannah, D.M. 2010. Water Sources and Habitat of Alpine Streams. U. Bundi (ed.), Alpine Waters, Hdb Env Chem, 6: 175–191.

Milner, A.M. and Petts, G.E. 1994. Glacial rivers: physical habitat and ecology. Freshw Biol, 32(2): 295–307 doi:10.1111/j.1365–2427.1994.tb01127.x

Minshall, G.W. and Minshall, J.N. 1977. Microdistribution of benthic invertebrates in a rocky mountain (U.S.A.) stream. *Hydrobiologia* 55: 231–249.

Minshall, G.W. 1984. Aquatic insect – substratum relationships. In *The Ecology of Aquatic Insects*, Resh VH, Rosenberg DM (eds). Praeger Scientific: New York; 358–400.

Morse, J.C., Yang, L.F. and Tian, L.X. 1994. Aquatic Insects of China Useful for Monitoring Water Quality. Nanjing: Hohai University Press, 1–570.

Newman, R.M. 1991. Herbivory and detritivory on freshwater macrophytes by invertebrates: a review. Journal of the North American Benthological Society, 10(2): 89–114.

Obrdlik, P. and Fuchs, U. 1991. Surface water connection and the macrozoobenthos of two types of floodplains on the Upper Rhine. *Regulated Rivers: Research and Management* 6: 279–288.

Paillex, A., Castlella, E. and Carron, G. 2007. Aquatic macroinvertebrate response along a gradient of lateral connectivity in river floodplain channels. *Journal of the North American Benthological Society* 26(4): 779–796.

Pan, B.Z., Wang, H.J., Liang, X.M. and Wang, H.Z. 2009. Factors influencing chlorophyll a concentration in the Yangtze-connected lakes. Fresenius Environmental Bulletin, 18(10): 1894–1900.

Pan, B.Z., Wang, H.J., Liang, X.M. and Wang, H.Z. 2011. Macrozoobenthos in Yangtze flood-plain lakes: patterns of density, biomass and production in relation to river connectivity. Journal of the North American Benthological Society, 30 (2): 589–602.

Pan, B.Z., Wang, Z.Y. and Xu, M.Z. 2012. Macroinvertebrates in abandoned channels: assemblage characteristics and their indications for channel management. River Research and Applications, 28(8): 1149–1160.

Pan, B.Z., Wang, H.J., Liang, X.M., Wang, Z.X., Shu, F.Y. and Wang, H.Z. 2008. Macro-zoobenthos in Yangtze oxbows: community characteristics and courses of resources decline. *Journal of Lake Science* 20(6): 806–813 (in Chinese with English abstract).

Plafkin, J.L., Barbour, M.T., Porter, K.D., Gross, S.K. and Hughes, R.M. 1989. Rapid bioassessment protocols for use in streams and rivers: Benthic macroinvertebrates and fish. U. S. Environmental Protection Agency Office of Water EPA/444/4-89-001, Washington, 1–170

Qian, J., Wang, G.X., Ding, Y.J. and Liu, S. 2006. The land ecological evolutional patterns in the source areas of the Yangtze and Yellow Rivers in the past 15 years, China. Environmental Monitoring and Assessment, 116: 137–156.

Quinn, J.M. and Hickey, C.W. 1990. Magnitude of effects of substrate particle size, recent flooding, and catchment development on benthic invertebrates in 88 New Zealand rivers. New Zealand Journal of Marine and Freshwater Research, 24: 411–427.

Richardot-Goulet, M., Castella, E. and Castella, C. 1987. Clarification and succession of former channels of the French upper Rhone alluvial plain using mollusca. *Regulated Rivers: Research and Management* 1: 111–127.

Rooke, B.J. 1984. The invertebrate fauna of four macrophytes in a lotic system. Freshwater Biology, 14(5): 507–513.

Rosine, W.N. 1955. The distribution of invertebrates on submerged aquatic plant surfaces in Muskee Lake, Colorado. Ecology, 36(2): 308–314.

Scheffer, M. 2004. Ecology of Shallow Lakes. Kluwer Dordrecht: Academic Publishers.

Schwarz, W.L., Malanson, G.P. and Weirich, F.H. 1996. Effect of landscape position on the sediment chemistry of abandoned-channel wetlands. *Landscape Ecology* 11(1): 27–38.

Sharma, R.C. and Rawat, J.S. 2009. Monitoring of Aquatic Macroinvertebrates as Bioindicator for Assessing the Health of Wetlands: A Case Study in the Central Himalayas, India. Ecological Indicators, 9: 118–28.

Sheng, H.Y., Yang, G.H., Wang, D.X., Fen, Y.Z. and Ren, G.X. 2007. Current status of grassland resources utilization and the corresponding solutions in a sustainable development context in the regions of three river source. Journal of Northwest Sci-Tech University of Agriculture and Forest, 35(7): 147–153. (in Chinese)

Shi, X.H., Qin, N.S., Xu, W.J., Yang, G.L., Liu, Q.C., Yan, H.Y., Wang, Q.C. and Feng, S.Q. 2007. The variations characteristic of the runoff in the source regions of the Yangtze River from 1956 to 2004. Journal of Mountain Science, 25(5): 513–523. (in Chinese)

Skjelkvåle, B.L. and Wright, R.F. 1998. Mountain lakes; sensitivity to acid deposition and global climate change. Ambio, 27: 280–286.

Smith, M.J., Kay, W.R., Edward, D.H.D., Papas, J. and Richardson, K.S.J. 1999. Using macroinvertebrates to assess ecological condition of rivers in Western Australia. Freshwater Biology, 41: 269–282.

Smith, M.J., Kay, W.R., Edward, D.H.D., Papas, P.J., Richardson, K.St.J., Simpson, J.C., Pinder, A.M., Cale, D.J., Horwitz, P.H.J., Davis, J.A., Yung, F.H., Norris, R.H. and Halse, S.A. 1999. AusRivAS: Using macroinvertebrates to assess ecological condition of rivers in Western Australia. *Freshwater Biology* 41: 269–282.

Smock, L.A. 1983. The influence of feeding habits on whole-body metal concentrations in aquatic insects. Freshw Biol, 13(4): 301–311 doi:10.1111/j.1365-2427.1983.tb00682.x

Song, M.H., Ma, Y.M., Zhang, Yu, Li, M.S., Ma, W.Q. and Sun, F.L. 2011. Analyses of characteriatics and trend of air temperature variation along the Brahmaputra valley. Climatic and environmental research, 16(6): 760–766 (in Chinese)

Strayer, D. 1985. The benthic micrometazoans of Mirror Lake, New Hampshire. Archiv für Hydrobiologie 72: 287–426.

Sun, J.G., Li, B.G. and Lu, Q. 2005. Modeling grassland productivity based on DEM in Gonghe basin of Qinghai Province. Resource Science, 27(4): 44–49.

Tang, X.P. 2003. Basic ecological characteristics of the Three-Rivers Source Area and design of the nature reserve. Forest Resources Management, (1): 38–44 (in Chinese with English abstract).

Tarr, T.L., Baber, M.J. and Babbitt, K.J. 2005. Macroinvertebrate community structure across a wetland hydroperiod gradient in southern New Hampshire, USA. Wetlands Ecology and Management, 13(3): 321–334.

Ter Braak, C.J.F. and Šmilauer, P. 2002. CANOCO Reference Manual and Users Guide to Canoco for Windows. Software for Canonical Community Ordination (Version 4.5). New York: Microcomputer Power, 1–500

Tockner, K., Schiemer, F., Baumgartner, C., Kum, G., Weigand, E., Zweimüller, I. and Ward, J.V. 1999. The Danube restoration project: species diversity patterns across connectivity gradients in the floodplain system. *Regulated Rivers: Research and Management* 15: 245–258.

Tomanova, S., Tedesco, P.A., Campero, M., Van Damme, P.A., Moya, N. and Oberdorff, T. 2007. Longitudinal and altitudinal changes of macroinvertebrate functional feeding groups in neotropical streams: a test of the River Continuum Concept. Fundamental and Applied Limnology, 170(3): 233–241 (Archiv für Hydrobiologie) doi:10.1127/1863-9135/2007/0170-0233

Van den Brink, F.W.B., Beljaards, M.J., Boots, N.C.A. and Van den Velde, G. 1994. Macro-zoobenthos abundance and community composition in three lower Rhine floodplain lakes with varying inundation regimes. *Regulated Rivers: Research and Management* 9: 279–293.

Van den Brink, F.W.B. and Van den Velde, G. 1991. Macrozoobenthos of floodplain waters of the rivers Rhine and Meuse in the Netherlands: a structural and functional analysis in relation to hydrology. *Regulated Rivers: Research and Management* 6: 265–277.

Verdonschot, P.F.M. 2001. Hydrology and substrates: determinants of oligochaete distribution in lowland streams (The Netherlands). *Hydrobiologia* 463: 249–262.

Wang, G.X., Li, N. and Hu, H.C. 2009. Hydrologic effect of ecosystem responses to climatic change in the source regions of Yangtze River and Yellow River. Advances in Climate Change Research, 5(4): 202–208. (in Chinese)

Wang, H.J., Pan, B.Z., Liang, X.M. and Wang, H.Z. 2006. Gastropods on submerged macro-phytes in Yangtze Lakes: community characteristics and empirical modelling. International Review of Hydrobiology, 91 (6): 521–538.

Wang, Z.Y., Lee, J.H.W., Cheng, D. and Duan, X. 2008. Benthic invertebrates investigation in the East River and habitat restoration strategies. Journal of Hydro-environment Research, 2: 19–27.

Wang, Z.Y., Lee, J.H.W. and Melching, C.S. 2012. Integrated River Training and Management. Beijing: Tsinghua University, 1–800.

Wang, Z.Y., Melching, C.S., Duan, X.H. and Yu, G. 2009. Ecological and Hydraulic Studies of Step-Pool Systems. J Hydraul Eng, 135(9): 705–717 doi:10.1061/(ASCE)0733-9429(2009)135:9(705)

Wang, Z.Y., Lee, J.H.W., Joseph, H.W., Cheng, D.S. and Duan, X.H. 2008. Benthic invertebrates investigation in the East River and habitat restoration strategies. *Journal of Hydro-environment Research* 2: 19–27.

Wang, Z.Y., Pan, B.Z., Xu, M.Z. and Yu, G.A. 2010. Preliminary investigation of aquatic ecology of the source region of the Yellow River. In *Landscape and Environment Science and Management in the Sanjiangyuan Region*, Brierley G, Li XL, Chen G (eds). Qinghai People's Publishing House: Xining; 93–104.

Ward, J.V. and Stanford, J.A. 1995. The serial discontinuity concept: extending the model to floodplain rivers. *Regulated Rivers: Research and Management* 10: 159–168.

Ward, J.V., Tockner, K. and Schiemer, F. 1999. Biodiversity of floodplain river ecosytems: ecotones and connectivity. Regulated Rivers: Research and Management 15: 125–139.

Ward, J.V. 1998. Riverine landscapes: biodiversity patterns, disturbance regimes, and aquatic conservation. *Biological Conservation* 83: 269–278.

Water and Waste Water Monitoring and Analysis Method Committee, 2002. Water and Waste Water Monitoring and Analysis Method (4th edition). Beijing: China Environmental Science Press. (in Chinese).

Waters, T.F. 1995. *Sediment in Streams: Sources, Biological Effects and Control*. American Fisheries Society: Bethesda.

Wise, D.H. and Molles, M.C. 1979. Colonization of artificial substrates by stream insects: influence of substrate size and diversity. *Hydrobiologia* 65: 69–74.

Wood, P.J. and Armitage, P.D. 1997. Biological effects of fine sediment in the lotic environment. *Environmental Management* 21: 203–217.

Wu, X., Shen, Z.Y., Liu, R.M. and Ding, X. 2008. Land use/cover dynamics in response to changes in environmental and socio-political forces in the upper reaches of the Yangtze River, China. Sensors, 8: 8104–8122.

Xie, Z.C., Liang, Y.L., Wang, J. and Feng, W.S. 1999. Preliminary studies of macroinvertebrates of the mainstream of the Changjiang (Yangtze) River. Acta Hydrobiologica Sinica, 23(Suppl.): 148–157.

Yan, Y.J. and Liang, Y.L. 1999. A study of dry-to-wet weight ratio of aquatic macroinvertebrates. Journal of Huazhong University of Science & Technology (Natural Science Edition), 27(9): 61–63. (in Chinese)

Yang, Z.P., Gao, J.X., Zhou, C.P., Shi, C.P., Zhao, P.L., Shen, W.S. and Hua, O.Y. 2011. Spatio-temporal changes of NDVI and its relation with climatic variables in the source regions of the Yangtze and Yellow rivers. Journal of Geographical Sciences, 21(6): 979–993.

You, Q.L., Kang, S.C., Wu, Y.H. and Yan, Y.P. 2007. Climate change over the Yarlung Zangbo River Basin during 1961–2005. *Journal of Geographical Sciences*, 17(4): 409–420 doi:10.1007/s11442-007-0409-y

Yu, G., Liu, L., Li, Z. et al., 2013. Spatial variability in valley setting and fluvial morphology in the source region of the Yangtze and Yellow Rivers. Journal of Geographical Sciences, 23(5): 817–832.

Zhang, S.F., Hua, D., Meng, X.J. and Zhang, Y.Y. 2011. Climate change and its driving effect on the runoff in the "Three-River Headwaters" region. Journal of Geographical Sciences, 21(6): 963–978.

Zhang, S.F., Jia, S.F., Liu, C.M., Cao, W.B., Hao, F.H., Liu, J.Y. and Yan, H.Y. 2004. Study on the changes of water cycle and its impacts in the source region of the Yellow River. Science in China (Series E), 47(suppl.): 142–151.

Zhao, W.H. and Liu, X.Q. 2010. Preliminary study on macrozoobenthos in Yarlung River River and its branches around Xiongcun, Tibet, China. Resources and Environment in the Yangtze Basin, 19(3): 281–286 (in Chinese)

Subject index

Printed and bound by CPI Group (UK) Ltd, Croydon, CR0 4YY

24/10/2024

01778285-0001